Lecture Notes in Mathematics 1590

Editors:
A. Dold, Heidelberg
B. Eckmann, Zürich
F. Takens, Groningen

Paolo M. Soardi

Potential Theory
on Infinite Networks

Springer-Verlag
Berlin Heidelberg New York
London Paris Tokyo
Hong Kong Barcelona
Budapest

Author

Paolo M. Soardi
Dipartimento di Matematica
Università di Milano
Via Saldini, 50
20133 Milano, Italy

E-Mail: soardi@vmimat.mat.unimi.it

Mathematics Subject Classification (1991): 05C, 31C, 43A, 52C, 60B, 60J, 94C

ISBN 3-540-58448-X Springer-Verlag Berlin Heidelberg New York

CIP-Data applied for

© Springer-Verlag Berlin Heidelberg 1994
Printed in Germany

Typesetting: Camera ready by author
SPIN: 10130158   46/3140-543210 - Printed on acid-free paper

CONTENTS

INTRODUCTION

While there is a large body of mathematical literature devoted to finite electrical networks, infinite networks have received growing attention only in the last two decades. The reason for this attention is probably the increasing interest for the discrete methods in all branches of mathematics. Actually, there is strong indication that, at least from the point of view of potential theory, infinite networks are a discrete model of noncompact Riemann manifolds. This is true in two ways: on one hand results in Riemann manifolds classification theory have been obtained by discretizing the manifold and then arguing on the network thus obtained. On the other hand, there are several results in potential theory on infinite networks which are the discrete counterpart of results on Riemann manifolds.

From an historical point of view, the infinite network which was first investigated by mathematicians (more than sixty years ago) is the infinite grid $\mathbb{Z}^n$ (especially for $n \leq 3$). The interest in this network was motivated by the study of the discretized Laplace or Poisson equation in the plane or in the space. The operator obtained by discretizing the laplacian is nothing else but the operator associated with the simple random walk on $\mathbb{Z}^n$, the one which assigns probability $1/2n$ of moving one step in each direction.

As in the case of $\mathbb{Z}^n$, every infinite network is associated with an irreducible, reversible Markov chain on the underlying graph. There is a discrete analogue of the laplacian on every network, and solving the corresponding discrete Poisson's equation is equivalent to solving Kirchhoff's equations. The converse is also true: every denumerable, reversible and irreducible Markov chain is the Markov chain associated with an electrical network. Thus there is an interplay between infinite networks and Markov chains theory: electrical concepts and results are explained in terms of probability theory and results on Markov chains are deduced from the laws governing electrical networks.

In these notes we tried to give a unified (but by no means comprehensive) approach to new developments in the theory of infinite networks. We start by describing the abstract model of a network and by formulating Kirchhoff's equations. Even though we point out the relationship with Markov chains and classify networks into transient and recurrent (according to the type of the associated Markov chain), chapters I and II are mainly devoted to exploring the model and to studying the elementary properties of currents, resistances etc.

Chapter III contains the basic results on infinite networks, existence theorems and Yamasaki's potential theory. We chose to confine our investigation to currents and potentials of finite energy. It should be noticed that this restriction does not result into loss of complexity. On the contrary, focusing our attention on the finite energy case allows us to introduce a machinery analogous to potential theory for

Riemann manifolds. This is still clearer in chapter VI, where we define and study the Royden compactification of a network along the lines of Royden's compactification for Riemann manifolds. Most definitions and results can be translated from the continuous to the discrete case. There are differences however, and we stress all the discrepancies and the results with no continuous analogue.

Chapter IV is a devoted to the problem of uniqueness. This is the part of the book where graph theoretical properties play a prominent role. To give some instances, the structure of the automorphism group, the number of ends, the rate of growth of balls, the relative size of the combinatorial boundary of finite sets are relevant to establish whether a given current input determines a unique current of finite energy in an infinite network.

In chapter V we apply the theory of the preceding chapters to study some particular networks. Using elementary Fourier analysis, we determine all possible potentials of finite energy for the n-dimensional (uniform) grid. Then we turn our attention to another remarkable class of networks i.e., the infinite cascades, both grounded and ungrounded. Two sections are devoted to trees, and, in particular, to the beautiful characterization of transient trees due to Benjamini and Peres. Finally, we examine edge graphs of tilings of the plane, a subject already introduced in chapter IV where edge graphs of tilings of the hyperbolic disk are considered.

The final chapter VII is devoted to rough isometries, a class of mappings between metric spaces (in particular graphs and Riemann manifolds) which, as pointed out by Kanai, preserve important properties such as parabolicity, isoperimetric inequalities etc.

Even though such investigations are outside the scope of these notes, let us mention that rough isometries provide the link between Riemann manifolds and network theory. In fact, under suitable assumptions, a noncompact Riemann manifold is roughly isometric to certain infinite graphs (ϵ-nets) whose vertices are points of the manifold and whose edges are geodesic arcs.

These notes were inspired by the tutorial book by Doyle and Snell (1984) on the relationship between electrical networks and random walks. We tried to organize in a coherent treatment of infinite networks many results contained in recent papers scattered among several journals; other results presented here (e.g. the last sections of chapter VI) are original and appear here for the first time.

We must also mention the excellent monograph by Zemanian (1991)(a) on infinite electrical networks. The intersection with these notes is small, as Zemanian's attention is focused on different features of infinite networks.

Finally, we wish to thank all the colleagues who read various parts of the manuscript and made useful suggestions which have been incorporated here. Special thanks are due to L. De Michele, G. Medolla, W. Woess and A. Zemanian.

CHAPTER I

KIRCHHOFF'S LAWS

§0. Examples of infinite electrical networks

It is known that there are several partial differential equations, such as Laplace's or Poisson's equations, whose finite difference models lead to electrical networks. If the domain is infinite, then the discretized model can be realized as an *infinite* electrical network. Let us present a concrete example studied in detail by Zemanian and Subramanian (1983); see also Zemanian (1991)(a).

Consider Poisson's equation

(i) $$\nabla \cdot (\sigma \nabla \phi) = f$$

in the three dimensional lower half space. Such an equation arises, for instance, in geophysical explorations when it is assumed that earth is flat and has conductivity $\sigma > 0$. To make the actual computation of the solution easier, in the flat earth model one supposes usually that the function σ depends only on the depth below the surface, but we need not make such an assumption.

One injects (or extracts) a current with distribution f into earth's surface. The variation of the potential ϕ along the surface is used to prognosticate on the variation of σ inside the earth, and the latter suggests the possible location of mineral ores.

The whole process just described requires solving equation (i) for given σ and f. One method of studying the solution ϕ consists in replacing (i) by a finite difference approximation. First order derivatives are replaced by first order differences, divided by the chosen increment of the variable Δx_k, $k = 1, 2, 3$, and second order derivatives by second order symmetric differences divided by the squared increment of the variable. The lower half space is replaced by an infinite semi–grid whose nodes have coordinates

(ii) $$(j_1 \Delta x_1, j_2 \Delta x_2, j_3 \Delta x_3)$$

where j_k, $k = 1, 2, 3$ is an integer and $j_3 \leq 0$.

To solve for the discrete values of ϕ a common procedure consists in truncating the grid along a large finite boundary. An alternative approach, which avoids the problem of specifying the appropriate boundary conditions for the truncated grid, is to describe the infinite grid as a purely conductive (resistive) infinite electrical network. The resulting network is described below.

The nodes (ii) of the grid are denoted simply by $v = (j_1, j_2, j_3)$. We orient the x_3 axis downward, so that $j_3 \geq 0$ always. The current is actually injected in the earth through current probes located at the surface. We represent this as the application of current sources $f = f(j_1, j_2, j_3)$, which have non–zero value only if $j_3 = 0$.

Denoting by e_k the unit vectors of the axes, every edge is of the form $[v, v + e_k]$. The conductance of the edge $[v, v + e_k]$ is defined as $c(v, v + e_k) = (\Delta x_k)^{-2} \sigma(v + e_k)$ (or also $(\Delta x_k)^{-2} \sigma(v + \frac{1}{2} e_k)$) and we complete the the definition of c in the upper half space by setting $c(v, v + e) = 0$ if $j_3 < 0$. Setting $U(v, v + e_k) = \phi(v + e_k) - \phi(v)$ the discretized form of equation (i) becomes

$$\sum_{k=1}^{3} c(v, v + e_k) U(v, v + e_k) - \sum_{k=1}^{3} c(v - e_k, v) U(v, v - e_k) = f(v).$$

Letting $i(v, v + e_k)$ denote the current flowing in $[v, v + e_k]$ and taking Ohms' law into account, the above equation becomes Kirchhoff's node equation

$$\sum_{k=1}^{3} i(v, v + e_k) - \sum_{k=1}^{3} i(v, v - e_k) = f(v).$$

. Another motivation for the study of infinite electrical networks arises from probability theory. We recall that a Markov chain, represented by a stochastic (infinite) matrix $\{p(x, y)\}$, is irreducible if there is positive probability of passing from any state x to any state y in a finite number of steps. The chain is called reversible if there is a positive measure (in general with infinite mass) $c(x)$ on the state space such that $c(x)p(x, y) = c(y)p(y, x)$ for all x and y. Reversibility means essentially that the process looks the same run backwards in time as run forwards in time.

Suppose we are given an irreducible, reversible Markov chain with countable state space V and transition probabilities $p(x, y)$. To such a chain we associate a graph Γ in the obvious way: the vertices of Γ are just the states in V and we say that there is an edge between x and y if $p(x, y)$ is positive. The resulting graph is countable and connected.

The conductance of the edge $[x, y]$ is defined to be $c(x)p(x, y)$. We will see later that this model is two–way: to any electrical network it is possible to associate an irreducible reversible Markov chain.

Given an infinite irreducible Markov chain, the problem arises whether the chain is transient or recurrent. This problem can be formulated in electrical terms.

Let $H(x,y)$ denote the probability of visiting y starting from x and let $G(x,y)$ denote the expected number of visits to y starting from x. From elementary Markov chain theory we know that the function H satisfies the equation

(iii)
$$\sum_z p(x,z)\big(H(x,y) - H(z,y)\big) = \Big(1 - \sum_z p(y,z)H(z,y)\Big)\delta(x,y)$$
$$= G(x,y)^{-1}\delta(x,y)$$

where δ is Kronecker's symbol. Now suppose that the associated network is energized by a 1 ampère current generator between infinity and y. Kirchhoff's node equation at x gives

(iv)
$$\sum_z i(x,z) = \delta(x,y)$$

where $i(x,z)$ is the current flowing in the edge $[x,z]$, $i(x,z) = -i(z,x)$, and the sum is over all z joined to x by an edge. Let $u(x)$ denote the corresponding potential at x. By Ohm's law equation (iv) becomes

(v)
$$\sum_z c(x,z)(u(x) - u(z)) = \delta(x,y).$$

Comparing equations (iii) and (v) we see that $c(x)^{-1}G(y,y)H(x,y)$ is a potential of the network (but not necessarily the unique potential, as we shall see). The effective resistance R between y and infinity is $c(y)^{-1}G(y,y)$. Thus the Markov chain is recurrent if and only if the effective resistance between a point and infinity is $+\infty$.

§1. Graphs and networks

A graph Γ consists of a (nonempty) set V, whose elements are called vertices (or points, or nodes) and a (nonempty) set $Y \subseteq V \times V$ of ordered pairs of vertices such that $[x,y] \in Y$ if and only if $[y,x] \in Y$. The elements of Y are called (oriented) edges or branches and are denoted by $[x,y]$, or xy, or simply by B (branch). We will write: $\Gamma = (V,Y)$. Thus V is the vertex set and Y the edge set of Γ.

The vertices x and y of $B = [x,y]$ are called the endpoints or the extremities of B. The vertex x is the initial point and the vertex y the terminal point.

In practice, a graph is often represented by a diagram, using the convention that a point marked on the diagram represents a vertex and a line joining two vertices x

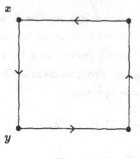

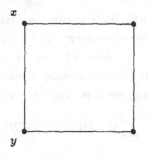

Fig. 1.1   Fig. 1.2

and y represents both the edges $[x,y]$ and $[y,x]$ (Figure 1.2). If we want to represent only the oriented edge $[x,y]$, then we join x to y by an arrow.

If the initial and terminal points of an edge B coincide i.e., $B = [x,x]$, then the edge is called a self–loop. A subset X of Y containing all the self–loops and such that, for $x \neq y$, $[x,y] \in X$ if and only if $[y,x] \notin X$ is called an orientation of Γ (Figure 1.1).

A subgraph of Γ is a graph $\Gamma' = (V', Y')$ such that $V' \subseteq V$ and $Y' \subseteq Y$ We will also write $\Gamma' \subseteq \Gamma$. A subgraph $\Gamma' \subseteq \Gamma$ is called induced by V' if, for every $x \in V'$ and $y \in V'$, the edge $[x,y]$ belongs to Y' whenever it belongs to Y.

Suppose that $\Gamma_i = (V_i, Y_i)$ is a subgraph of $\Gamma = (V, Y)$ for all $i \in I$. If $V = \cup_{i \in I} V_i$ and $Y = \cup_{i \in I} Y_i$, then we will say that Γ is the union of the graphs Γ_i and we will write $\Gamma = \cup_{i \in I} \Gamma_i$.

An exhaustion of Γ is a sequence of finite subgraphs $\Gamma_n = (V_n, Y_n)$ $(n = 1, 2, \dots)$ such that $\Gamma_n \subseteq \Gamma_{n+1}$ and $\Gamma = \cup_{n=1}^{\infty} \Gamma_n$.

We say that the extremities of an edge B belong to B, or that B is incident to its endpoints. If two edges are incident to the same vertex, they are called adjacent. If two vertices x, y are the two endpoints of the same edge, then we say that they are neighbours, and we write $x \sim y$. Note that if $[x,x] \in Y$, then $x \sim x$. The set of all neighbors of x will be always denoted by $V(x)$. The degree of x, denoted by $\deg(x)$, is the cardinality of $V(x)$. If $\deg(x) < \infty$ for all $x \in V$, then Γ is called locally finite.

DEFINITION. Suppose that $\Gamma_1 = (V_1, Y_1)$ and $\Gamma_2 = (V_2, Y_2)$ are graphs. Let $\phi : V_1 \mapsto V_2$ be a map such that $x \sim y$ implies $\phi(x) \sim \phi(y)$. Then, we say that ϕ induces a morphism from Γ_1 to Γ_2 or, by abuse of language, that ϕ is a morphism from Γ_1 to Γ_2.

To such a map ϕ we will always associate the map $\Phi = \phi \times \phi : Y_1 \mapsto Y_2$, defined by $\Phi([x,y]) = [\phi(x), \phi(y)]$. We say that the morphism is an isomorphism if

ϕ and Φ are bijections. In particular, an isomorphism of Γ onto itself is called an automorphism.

The automorphisms of a graph form a group denoted by Aut(Γ). A graph Γ is called vertex transitive if for every $x \in V$ and $y \in V$ there exists an automorphism such that $\phi(x) = y$. The graph Γ is called 1–transitive if for every $x \sim y$ and $x' \sim y'$ there exists an automorphism of the graph such that $\phi(x) = x'$ and $\phi(y) = y'$.

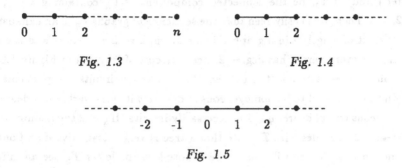

Fig. 1.3   Fig. 1.4

Fig. 1.5

DEFINITION. Let Γ be a graph. A path of length n in Γ is a subgraph isomorphic to the graph in Figure 1.3. A one–ended (or one–sided) infinite path is a subgraph isomorphic to the graph in Figure 1.4. A two–ended (or two–sided) infinite path is a subgraph isomorphic to the graph in Figure 1.5.

A two–ended path is also denoted by $\mathbb{Z}$, a one–ended path by $\mathbb{Z}_+$. The vertex x_0 corresponding to 0 in a path is called the initial point. We also say that the path starts at x_0. If the path has length n, then we say that the vertex x_n corresponding to n is the terminal point of the path. We also say that the path joins x_0 and x_n. The graph Γ is said to be connected if any two vertices are joined by a path.

Assumption. *Unless otherwise stated all the graphs Γ considered in these notes will be connected.*

Suppose that $V' \subseteq V$ is nonempty. We will say that V' is connected if any two vertices of V' are joined by a path whose vertices are still in V'. This amounts to saying that the subgraph of Γ induced by V' is connected.

The (geodesic) distance $d(x,y)$ between two vertices x and y is the minimum n such that there exists a path of length n joining x and y (see Figure 1.2, where $d(x,y) = 1$). Note that a locally finite connected graph has at most countably many vertices and edges (there are only finitely many vertices at distance 1 from a fixed vertex x, only finitely many vertices at distance 2, ...). In general we will say that Γ is countable if V (and consequently Y) is countable.

Assumption. *We will always assume that Γ is at most countable.*

An exhaustion of Γ always exists if Γ is connected and countable. To see this, fix $v \in V$ and let $S_k = \{x \in V : d(x,v) = k\}$. Each S_k is at most countable so that for every k there exists an increasing sequence of finite sets $S_{k,j}$ such that $\bigcup_j S_{k,j} = S_k$. Now, define Γ_n^* as the subgraph of Γ induced by $\bigcup_{k=1}^n S_{k,n}$. This subgraph is not necessarily connected, so that we proceed as follows. Set $\Gamma_0 = \Gamma_0^*$ (which is trivially connected) and let Γ_n be the connected component of Γ_n^* containing Γ_{n-1}, for $n = 1, 2, \ldots$. Then, it is easily seen that the sequence of graphs Γ_n is an exhaustion of Γ. A circuit of length n in a graph Γ is a subgraph with n vertices and n edges and such that every vertex has degree 2. See a circuit of length 4 in Figure 1.2.

A graph is called a tree if it is connected and has no circuits. In particular, a locally finite tree is said to be homogeneous of degree q if every vertex has degree q. A homogeneous tree of degree $q > 1$ is necessarily infinite. If $q = 2$ the homogeneous tree of degree 2 coincides with $\mathbb{Z}$. Note that a tree is vertex transitive if and only if it is homogeneous. A tree will usually be denoted by the letter T. See an infinite locally finite nonhomogeneous tree in Figure 1.6 and a homogeneous tree of degree 3 in Figure 1.7. Figure 1.8 represents a countable non locally finite tree.

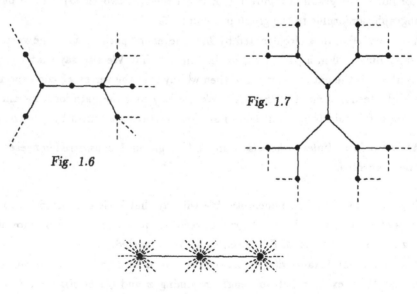

Fig. 1.6

Fig. 1.7

Fig. 1.8

Now we come to the definition of an electrical network.

DEFINITION. An electrical network (or, simply, a network) is a couple (Γ, r), where Γ is an at most countable connected graph and r is a positive function on Y

with the property that $r(x,y) = r(y,x)$ for all $x \sim y$ and

$$(1.1) \qquad c(x) = \sum_{y \in V(x)} \frac{1}{r(x,y)} < \infty \qquad \text{for all } x \in V.$$

Note that (1.1) is automatically satisfied in a locally finite graph. The value $r(x,y)$ of r at the edge $[x,y]$ is called the resistance of the edge; its reciprocal $c(x,y) = r(x,y)^{-1}$ is called the conductance of $[x,y]$. The sum $c(x)$ in (1.1) is called the total conductance at x.

REMARK. What we have in fact defined is a *resistive, linear* electrical network. In these notes we will always use the term "electrical network" or simply "network", to denote resistive, linear electrical network.

If $\Gamma' = (V', Y')$ is a subgraph of Γ and r' is the restriction of r to Y' we will say that (Γ', r') is a subnetwork of (Γ, r). If $V' = V$, we will say that (Γ', r') is obtained from (Γ, r) by *cutting* the edges in $Y \setminus Y'$.

Let (Γ_j, r_j), $j = 1, 2$, be two networks and let $\phi : V_1 \mapsto V_2$ be a graph isomorphism such that $r_2(\phi(x), \phi(y)) = r_1(x,y)$. Then, we say that ϕ is (or induces) a network isomorphism. If $(\Gamma_1, r_1) = (\Gamma_2, r_2)$, then we say that ϕ is an automorphism. More generally, we define a network morphism in the following way. Let (Γ_j, r_j) for $j = 1, 2$ be networks. Let $\phi : V_1 \mapsto V_2$ be a graph morphism. Suppose that

$$\frac{1}{r_2(B')} \geq \sum_{\{B \in \Phi^{-1}(B')\}} \frac{1}{r_1(B)}$$

whenever $B' \in \Phi(Y)$ is not a self-loop. Then we will say that ϕ is (or induces) a network morphism from (Γ_1, r_1) to (Γ_2, r_2).

Suppose for instance that (Γ, r) is a network and let $U \subseteq V$. Suppose that $\phi : V \mapsto V$ has the property that $\phi(x) = x$ if $x \in U$ and that there exists $b \notin U$ such that $\phi(x) = b$ for all $x \notin U$. Then ϕ induces a graph morphism between Γ and $\Gamma' = (V', Y')$, where $V' = \phi(V) = U \cup \{b\}$ and $Y' = \Phi(Y)$. More precisely, Y' consists of all the pairs $[x,y]$, where x and y are neighbors in U, and of all the pairs $[x,b]$ and $[b,x]$, where $x \in U$ and x has a neighbor not in U. The self-loop $[b,b]$ is in Y'.

For instance, let Γ be the graph in Figure 1.9. Let U be the set of all nodes x such that $d(x,o) < 3$. The resulting graph Γ' is represented in Figure 1.10. The edge joining z with b in Γ' is in fact obtained by identifying three edges of Γ with $[z,b]$ in Γ'. Note the self-loop at b.

For every edge $B' \in Y'$ which is not a self-loop define

$$r'(B') = \Big(\sum_{\{B \in Y : \Phi(B) = B'\}} \frac{1}{r(B)} \Big)^{-1}.$$

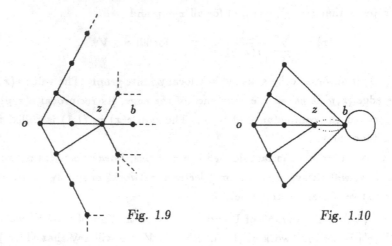

Fig. 1.9   Fig. 1.10

If B' is a self loop we assign to B' any value $r'(B')$. If $B' = [x, y]$, with $x \in U$ and $y \in U$, then $r'(B') = r(x, y)$. If $B' = [x, b]$ for some $x \in U$, then by (1.1) $0 < r'(B') < \infty$. Hence r' is a well defined resistance function on Y'.

If $x \in U$, then the total conductance at x is unchanged i.e., $c(x) = c'(x)$, while the total conductance $c'(b)$ at b may be ∞. However, if U is finite, then it is immediately seen that $c'(b) < \infty$.

Suppose that the resistance function r' satisfies condition (1.1). Then, we say that the network (Γ', r') just described is obtained from (Γ, r) by *shorting* together the vertices in $V \setminus U$.

Finally, an exhaustion of a network (Γ, r) is an increasing sequence of finite subnetworks (Γ_n, r_n), such that the sequence $\Gamma_n = (V_n, Y_n)$ is an exhaustion of Γ and r_n is the restriction of r to Y_n.

§2. Chains and cochains

The first rigorous analysis of infinite networks is due to H. Flanders (1971). The setup introduced below is essentially the same as in Flanders.

First of all, currents in electrical networks will be represented by 1–chains.

DEFINITION. A 1–chain on a graph Γ is a real valued function $I = \{i(x, y)\}$ defined on Y such that

$$i(x, y) = -i(y, x) \qquad \text{for every } x \sim y.$$

In particular, if only finitely many $i(x, y)$ are different from zero, we say that I is a finite 1–chain. Note that the definition of I implies that $i(x, y) = 0$ for every self–loop.

Fix an orientation X of Γ. A 1–chain is conveniently represented by a real linear combination

$$(1.2) \qquad\qquad I = \sum_{B \in X} i(B) \delta_B$$

where δ_B assigns value 1 to B and value 0 to the other edges in X. In the sequel we agree that, whenever $I, J, \dots, Z$ denote 1–chains, then the corresponding values on the edge B are denoted by $i(B), j(B), \dots, z(B)$.

External current sources will be represented by 0–chains.

DEFINITION. A 0–chain on Γ is a real valued function $\jmath$ defined on V. As in the case of 1–chains, every 0–chain $\jmath$ can be written as a linear combination of elementary functions. Namely, let δ_z be the function on V which assign value 1 to z and value 0 to the other vertices. A 0–chain $\jmath$ has the form

$$(1.3) \qquad\qquad \jmath = \sum_{z \in V} \jmath(z) \delta_z.$$

If $\jmath$ is finitely supported, then we say that $\jmath$ is a finite 0–chain. A finite 0–chain of the form $\jmath = \kappa(\delta_b - \delta_a)$, $a, b \in V$, κ real, is called a dipole (with poles a and b). The set of all 1–chains and the set of all 0–chains are endowed with the natural vector space structure on the real field.

Next we define the boundary operator from 1–chains to 0–chains.

DEFINITION. We denote by $\mathbf{C}$ the vector subspace of all 1–chains $I = \{i(x, y)\}$ on Γ such that

$$(1.4) \qquad\qquad \sum_{y \in V(x)} |i(x, y)| < \infty \qquad \text{for all } x \in V.$$

The boundary ∂I of a 1–chain $I \in \mathbf{C}$ is the 0–chain

$$(1.5) \qquad\qquad \partial I = \sum_{x \in V} \Big(\sum_{y \in V(x)} i(x, y) \Big) \delta_x.$$

The value $\sum_{y \in V(x)} i(x, y)$ of ∂I at x is denoted by $(\partial I)(x)$.

Clearly, finite 1–chains belong to $\mathbf{C}$. Moreover if Γ is locally finite, then every 1–chain is in $\mathbf{C}$.

DEFINITION. Let $\jmath$ be a finite 0–chain as in (1.3). The boundary $\partial \jmath$ of $\jmath$ is defined as the real number

$$\partial \jmath = \sum_{x \in V} \jmath(x).$$

It is immediately seen that for every finite 1–chain I $\partial \partial I = 0$.

DEFINITION. A 1–chain Z such that $\partial Z = 0$ is called a cycle.
For instance, the 1–chain which takes value 1 on the oriented edges of the graph in Figure 1.1 is a finite cycle.

DEFINITION. A linear functional on the vector subspace of all finite 1–chains on Γ is called a 1–cochain on Γ. Analogously, a linear functional on the vector subspace of all finite 0–chains is called a 0–cochain on Γ.

For every $B \in X$ let χ_B denote the functional on the linear subspace of the finite 1–chains such that $\chi_B(\delta_D) = 1$ if $B = D$ and $\chi_B(\delta_D) = 0$ if $B \neq D$, for every $D \in X$. Every 1–cochain E can be represented as a formal linear combinations of the elementary functionals χ_B, $B \in X$. Namely

$$E = \sum_{B \in X} e(B)\chi_B.$$

Denoting by $\langle \cdot, \cdot \rangle$ the duality between finite 1–chains and 1–cochains, we have for every 1–cochain E and every finite 1–chain $I = \sum_{B \in X} i(B)\delta_B$

$$\langle E, I \rangle = \sum_{B \in X} e(B)i(B).$$

Analogously, for every $x \in V$ let χ_x denote the 0–cochain such that $\chi_x(\delta_y) = 1$ if $x = y$ and 0 otherwise. Then, every 0–cochain u can be expressed in the form

(1.6) $$u = \sum_{x \in V} u(x)\chi_x.$$

Denoting by $\langle \cdot, \cdot \rangle$ the duality between finite 0–chains and 0–cochains, we have

$$\langle u, \jmath \rangle = \sum_{x \in V} u(x)\jmath(x).$$

Finite 1–cochains and finite 0–cochains are defined as functionals taking only finitely many nonzero values on the set of the elementary 1–chains χ_B and the elementary 0–chains χ_x, respectively.

0–cochains will represent potentials; the corresponding voltages will be represented by coboundaries.

DEFINITION. The coboundary of a 0–cochain u is the unique 1–cochain βu such that, for all finite 1–chains K,

(1.7) $$\langle \beta u, K \rangle = \langle u, \partial K \rangle.$$

The following results are also well known from elementary topology. We give the proofs for convenience of the reader.

Theorem (1.8). *Suppose that the 1–cochain E is a coboundary, and let $E = \beta u$. If u' is another 0–cochain such that $E = \beta u'$ then*

$$(1.9) \qquad\qquad u - u' = \text{const.} \sum_{x \in V} \chi_x.$$

Proof. By (1.7) and the linearity of β, $\beta u = \beta u'$ holds if and only if $0 = \langle \beta u - \beta u', B \rangle = \langle u - u', \partial(\delta_B) \rangle$ for all $B = [x, y] \in X$ i.e., if and only if

$$(1.10) \qquad\qquad \langle u - u', \delta_x \rangle = \langle u - u', \delta_y \rangle$$

for all $x \sim y$. By connectedness (1.10) holds for any couple of vertices of Γ. Hence $u(x) - u'(x) = \text{const.}$, so that (1.9) holds. $\square$

Theorem (1.11). *A 1–cochain E is a coboundary if and only if*

$$(1.12) \qquad\qquad \langle E, Z \rangle = 0$$

for every finite cycle Z.

Proof. In one direction (1.12) follows immediately from the definition of coboundary (1.7). Conversely, fix a node $o \in V$. By connectedness, for every $x \in V$ there is a finite 1–chain K_x such that $\partial K_x = \delta_x - \delta_o$. Let u denote the 0–cochain defined by $\langle u, \delta_x \rangle = \langle E, K_x \rangle$. By (1.12) u does not depend on the choice of K_x satisfying $\partial K_x = \delta_x - \delta_o$. Therefore, for all $x \in V$,

$$\langle \beta u, K_x \rangle = \langle u, \partial K_x \rangle = \langle u, \delta_x \rangle = \langle E, K_x \rangle.$$

Since the 1–chains K_x generate the subspace of finite 1–chains, $\beta u = E$. $\square$

Theorem (1.13). *A finite 0–chain $\jmath$ is the boundary of a finite 1–chain if and only if $\partial \jmath = 0$.*

Proof. We already observed that a boundary $\jmath$ satisfies $\partial \jmath = 0$. Conversely, if $\jmath$ is as in (1.3) and $o \in V$ is fixed

$$\jmath = \sum_{x \in V} \jmath(x) \delta_x - \left(\sum_{x \in V} \jmath(x) \right) \delta_o$$

$$= \partial \left(\sum_{x \in V} \jmath(x) K_x \right)$$

where K_x is a finite 1–chain such that $\partial K_x = \delta_x - \delta_o$. $\square$

It is convenient to interpret resistances as a single linear transformation from 1–chains to 1–cochains.

DEFINITION. The resistance operator $\mathcal{R}$ mapping 1–chains to 1–cochains is defined as

$$\mathcal{R}(I) = \sum_{B \in X} i(B)r(B)\chi_B$$

for every $I = \{i(B)\}$.

Note that $\mathcal{R}$ is invertible. The inverse operator is denoted by $\mathcal{R}^{-1}$. If 1–cochains are interpreted as voltages, then the resistance operator encompasses Ohms' law.

§3. Kirchhoff's equations

To complete the model of an electrical network we must give a formal expression to Kirchhoff's laws. Remember that Kirchhoff's node law asserts that the algebraic sum of all currents at every node x is zero, and Kirchhoff's loop law asserts that the voltage drop along every (finite) cycle is zero.

Suppose that a 1–cochain F and a 0–chain $\jmath$ are assigned. Kirchhoff's laws are formally expressed by the following equations in the unknown 1–chain $I \in \mathbf{C}$.

(1.14) $\partial I + \jmath = 0,$

(1.15) $\langle \mathcal{R}(I) - F, Z \rangle = 0$ for all finite cycles Z.

Equation (1.14) expresses the node law and equation (1.15) the loop law. If $\jmath = 0$ and $F = 0$, then the above equations are called homogeneous.

The 1–cochain F represents the "internal voltage generators" while the 0–chain $\jmath$ represents the "the external current sources". F and $\jmath$ will be called "the energy sources".

DEFINITION. A 1–chain $I \in \mathbf{C}$ satisfying equation (1.14) is called a flow generated by $\jmath$. If I satisfies both (1.14) and (1.15), then I is called a current generated by $\jmath$ and F.

Let F be the 1–cochain appearing in (1.15), and set $E = \mathcal{R}^{-1}(F)$. Suppose that $E \in \mathbf{C}$. Setting $\imath = \jmath + \partial E$, and calling I again the new unknown $I - E$, equations (1.14), (1.15) take the form

(1.16) $\partial I + \imath = 0,$

(1.17) $\langle \mathcal{R}(I), Z \rangle = 0$ for all finite cycles Z.

Analogously, we may transform Kirchhoff's equation (1.14), (1.15) in such a way as to eliminate the 0–chain $\jmath$. In order to do this we must first assume that $\jmath$ is the boundary of some 1–chain J. This assumption is natural, otherwise the equations can not be solved. For instance, if $\jmath$ is finite, then by Theorem (1.13) we must assume that $\partial \jmath = 0$.

Arguing as before we find that, denoting by I again the new unknown $I + J$ and by E the 1–cochain $F + \mathcal{R}(J)$, Kirchhoff's equations take the form

$$(1.18) \qquad \partial I = 0,$$

$$(1.19) \qquad \langle \mathcal{R}(I) - E, Z \rangle = 0 \qquad \text{for all finite cycles } Z.$$

Note that, when obtained from (1.14)–(1.15), equations (1.18)–(1.19) depend on the choice of J such that $\partial J = \jmath$.

In the sequel we will be mainly interested in solving equations (1.16), (1.17), since they are equivalent to the discrete version of Poisson's equation. Namely, suppose that $I \in \mathbf{C}$ is a solution of Kirchhoff's equations in the form (1.16), (1.17). By Theorem (1.11) $\mathcal{R}(I)$ is a coboundary, so that there is a 0–cochain $u = \sum_{x \in V} u(x) \chi_x$, unique up to a constant by Theorem (1.8), such that $\beta u = \mathcal{R}(I)$. By (1.7) and the definition of resistance operator in §2, this amounts to saying that for every edge $B = [x, y] \in X$

$$r(x,y)i(x,y) = \langle \beta u, \delta_B \rangle = \langle u, \partial(\delta_B) \rangle = u(x) - u(y).$$

Hence

$$(1.20) \qquad i(x,y) = c(x,y)(u(x) - u(y))$$

for every $x, y \in V$ such that $x \sim y$.

Let $\imath = \sum_{x \in V} \imath(x) \chi_x$. Substituting the value (1.20) of $i(x,y)$ into equation (1.16), we obtain that the function u must satisfy at every node x the equation

$$\sum_{y \in V(x)} c(x,y)(u(x) - u(y)) + \imath(x) = 0.$$

Taking (1.1) into account, we set for all $x, y \in V$

$$(1.21) \qquad p(x,y) = \begin{cases} c(x,y)c(x)^{-1}, & \text{if } x \sim y \\ 0 & \text{otherwise.} \end{cases}$$

Observe that $\sum_{y \in V} p(x,y) = 1$ for all $x \in V$. Since $I \in \mathbf{C}$, (1.4) implies

$$(1.22) \qquad \sum_{y \in V(x)} p(x,y)|u(y)| < \infty \qquad \text{for all } x \in V.$$

Defining $f(x) = -\imath(x)c(x)^{-1}$, we get

$$(1.23) \qquad u(x) - \sum_{y \in V} p(x,y)u(y) = f(x), \qquad \text{for all } x \in V.$$

Clearly, if u satisfies (1.22), (1.23), then also $u + \text{const.}$ does. We have proved the following theorem.

Theorem (1.24). *For every solution I of equations (1.16), (1.17) there is a function u, unique up to a constant, which satisfies (1.20), (1.22), and (1.23). Conversely, for every u satisfying (1.22) and (1.23), the I chain defined by (1.20) is a current in $\mathbf{C}$ satisfying (1.16), (1.17).*

DEFINITION. The function u associated with a current $I \in \mathbf{C}$ by (1.20), unique up to an additive constant, is called (the) potential of I. We will also use the notation: $I = du$.

Theorem (1.24) transforms the problem of solving Kirchhoff's equations (1.16), (1.17) into the problem of solving equation (1.23). By linearity, a solution I is unique provided the *homogeneous* Kirchhoff's equations admit only the trivial solution. Since in the homogeneous case $f = 0$, the current is unique if and only if every solution of the equation

$$u(x) - \sum_{y \in V} p(x,y)u(y) = 0, \quad \text{for all } x \in V,$$

is constant.

§4. The associated Markov chain

It is a standard procedure since the paper of Nash–Williams in (1959) to associate with a given network (Γ, r) a Markov chain with state space the vertex set of Γ and transition probabilities $p(x,y)$ as in (1.21). In fact we see from (1.21) that the (finite or infinite) matrix $P = \{p(x,y)\}$ is stochastic and defines a Markov chain, called the associated Markov chain.

If an initial probability distribution p is specified, then there exists a unique (up to equivalences) sequence of random variables $X_1, X_2, \ldots$, taking values in V, such that $\Pr(X_{n+1} = y | X_n = x) = p(x,y)$. As $p(x,y)$ represents the probability of passing in one step from x to y, we will also say that the associated Markov chain represents a (nearest neighbour) random walk on Γ. If Γ is locally finite and all the conductances are equal to 1, then the random walk is called simple.

Let P^n the (finite or infinite) matrix obtained by multiplying P n–times with itself. Let $p^n(x,y)$ denote the entries of this matrix. Clearly, $p^n(x,y) = \Pr(X_{m+n} = y | X_m = x)$. If $x \sim y$, then the probability $p(x,y)$ is always positive, as $c(x,y) > 0$ for every edge. Since Γ is connected, for any two vertices x and y we have $p^n(x,y) > 0$ for $n = d(x,y)$. Hence the chain is irreducible and all the states x belong to the same class. Furthermore

$$c(x)p(x,y) = c(y)p(y,x), \quad \text{for all } x \sim y.$$

Hence the chain is reversible.

Conversely, we have shown in §0 that every irreducible reversible Markov chain with countable state space is the associated Markov chain of an electrical network.

The matrix P acts by left multiplication on the linear space of all functions on V (identified with column vectors) satisfying (1.22). The corresponding operator is still denoted by P. We have

(1.25) $$P(u)(x) = \sum_{y \in V} p(x,y)u(y).$$

P is also called the random walk operator on Γ.

EXAMPLES (1.26). (1) Let V be the set of all n-dimensional vectors with integer coordinates. Let $x \sim y$ if and only if $x - y = \pm e_j$, for some j. Here $e_1, \ldots, e_n$ are the coordinate vectors. Set $r(B) = 1$ for all edges. Such a network is called the n-dimensional infinite grid and is denoted by $\mathbf{Z}^n$. See Figure 1.11 for $n = 2$.

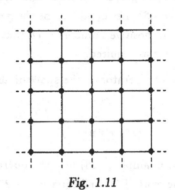

Fig. 1.11

The random walk operator on $\mathbf{Z}^n$ has the form

$$P(u)(x) = (2n)^{-1} \sum_{j=1}^{n} u(x + e_j) + (2n)^{-1} \sum_{j=1}^{n} u(x - e_j).$$

(2) Let $\Gamma = T_q$ be the homogeneous tree of degree q (see Figure 1.7). Let $r(B) = 1$. Then

$$P(u)(x) = q^{-1} \sum_{y \in V(x)} u(y).$$

(3) More generally, let Γ be any locally infinite graph and let $r(B) = 1$ for all edge B. Then

$$P(u)(x) = \frac{1}{\deg(x)} \sum_{y \in V(x)} u(y).$$

(4) Let Γ be the two-sided infinite path $\mathbb{Z}$ and assign conductance c_n to the edge $[n, n+1]$, $-\infty < n < +\infty$. Then

$$P(u)(n) = \frac{c_{n-1}}{c_{n-1} + c_n} u(n-1) + \frac{c_n}{c_{n-1} + c_n} u(n+1).$$

(5) Let G be a discrete countable group with a set of generators H and suppose that $H^{-1} = H$. The Cayley graph of G (with respect to H) is defined in the following way: the vertex set is G, and we say that $x \sim y$ if $y^{-1}x \in H$. Let μ be a nonnegative function on H such that $\sum_{h \in H} \mu(h) < \infty$ and $\mu(h) = \mu(h^{-1})$. For all $y \in G$ and $h \in H$ we set $r(y, yh) = 1/\mu(h)$. Thus r is a well defined resistance function on the edges of the Cayley graph of G.

The network of example (2) can obtained in this way: here G is the free product of q copies of the group with 2 elements, $\mathbb{Z}_2^{(j)} = \{1, a_j : a_j^2 = 1\}$, $j = 1, \ldots, q$, and H is the set $\{a_1, a_2, \ldots, a_q\}$. The function μ assigns value 1 to every generator.

Another important example is provided by the free product of infinitely many copies of $\mathbb{Z}_2$. The resulting Cayley graph is the countable tree each of whose vertices has infinite degree (see Figure 1.8). Let $a_1, a_2, \ldots$ be the generators of the copies of $\mathbb{Z}_2$ and let μ_j be a sequence of positive numbers such that $\sum_{j=1}^{\infty} \mu_j < \infty$. Setting $\mu(a_j) = \mu_j$ defines a function μ as required.

Now let 1 denote the identity operator on the space of all functions on V. Equation (1.23) becomes

(1.27) $(1 - P)(u) = f.$

DEFINITION. On account of Example (1.26) (1), the operator $1 - P$ is called the (discrete) Laplacian of the network (Γ, r) and equation (1.27) is called the (discrete) Poisson equation.

Now we give a short description of the probabilistic quantities which will be relevant in the following. We refer to the book by Kemeny, Snell and Knapp (1966) for the proofs of the statements below.

Let $X_1, X_2, \ldots$ be a sequence of random variables as above. We set, for all $x, y \in V$,

$$H(x, y) = \sum_{n=0}^{\infty} \Pr(X_n = y, X_j \neq y, 0 \le j < n | X_0 = x),$$

$$\overline{H}(x, y) = \sum_{n=1}^{\infty} \Pr(X_n = y, X_j \neq y, 0 < j < n | X_0 = x).$$

$H(x, y)$ and $\overline{H}(x, y)$ represent the probability of reaching y starting from x (including or excluding, respectively, the time $n = 0$).

Suppose $x, y \in V$. We set

$$G(x,y) = \sum_{n=0}^{\infty} p^n(x,y).$$

$G(x,y)$ represents the expected number of visits to y when the random walk starts at x. The following relations between G, H and $\overline{H}$ are well known.

$$G(x,y) = \delta(x,y) + \overline{H}(x,y)G(y,y),$$

$$\sum_{z} p(x,z)H(z,y) = \overline{H}(x,y),$$

(1.28)

$$G(x,y) = H(x,y) \sum_{n=0}^{\infty} \left(\overline{H}(y,y)\right)^n.$$

Here $\delta(x,y)$ is Kronecker's delta. Note that $G(x,y)$ is finite if and only if $\overline{H}(y,y) < 1$.

The (irreducible, reversible) Markov chain $\{p(x,y)\}$ is called transient if there exists $x \in V$ such that $\overline{H}(x,x) < 1$. Otherwise it is called recurrent. If $\overline{H}(x,x) = 1$ for some x, then $H(x,y) = 1$ for every x and y, since there is positive probability of reaching any state x from any other state y. Therefore $G(x,y)$ is finite for all x, y if and only if the random walk is transient.

DEFINITION. A network (Γ, r) will be called transient or recurrent depending on whether the associated Markov chain is transient or recurrent. When all the resistances are equal to 1 we will also say that the graph Γ is transient or recurrent.

The quantity $G(x,y)$ is the (x,y)-entry of the matrix $G = \sum_{n=0}^{\infty} P^n$. If the Markov chain is transient, then

(1.29) $$\sum_{z \in V} G(x,z)p(z,y) = \sum_{z \in V} p(x,z)G(z,y) = G(x,y) - \delta(x,y).$$

Suppose that (Γ, r) is a transient network. We set, for every function f on V,

(1.30) $$G(f)(x) = \sum_{y \in V} G(x,y)f(y)$$

whenever this expression is well defined (possibly ∞). Because of (1.29) the operator defined in (1.30) is also called the Green operator and $G(x,y)$ the Green function of P.

Let $b \in V$ be a fixed vertex. We form a new Markov chain $\{p(x,y;b)\}$ by making b absorbing i.e., $p(x,y;b) = p(x,y)$ if $x \neq b$, $p(b,b;b) = 1$, $p(b,y;b) = 0$ if $y \neq b$. Accordingly, we set for every x

$$G(x,y;b) = \begin{cases} \sum_{n=0}^{\infty} p^n(x,y;b) & \text{if } y \neq b \\ 0 & \text{if } y = b. \end{cases}$$

$G(x, y; b)$ is the expected number of visits to y before hitting b, starting from x. If $y \neq b$, then $G(x, y; b)$ coincides with the Green function of the Markov chain $\{p(x, y; b)\}$.

Note that $G(x, y; b)$ is always finite. This is clear if the chain is transient, as $p^n(x, y) \geq p^n(x, y; b)$. If the chain is recurrent, for every $x \neq b$ denote by $H(x, y; b)$ and $\overline{H}(x, y; b)$ the probability of visiting y before b when the process starts at x (including or excluding respectively the time $n = 0$). Clearly $H(x, y; b) \leq H(x, y)$ and $\overline{H}(x, y; b) \leq \overline{H}(x, y)$. The analogue of the second equation of (1.28) implies for $x \neq b$

$$(1.31) \qquad \overline{H}(x, x; b) = \sum_{z \neq b} p(x, z) H(z, x; b)$$
$$< 1.$$

because $H(z, x; b) < 1$ for at least one z. By (1.31), and by the analogue, for random walks with an absorbing state, of the third and first equations of (1.28), $G(x, x; b) < \infty$ and $G(x, y; b) < \infty$.

§5. Harmonic functions

Suppose that equation (1.27) has at least one solution. Such a solution is unique up to an additive constant if and only if the homogeneous equation $(1 - P)(u) = 0$ has no nonconstant solutions. This is not the case in general, as Example (1.34) below shows. In Chapter IV we will be interested in the uniqueness of the solutions of (1.27) in a certain Hilbert space. In this section we confine ourselves to establishing some general properties of discrete harmonic functions. We start with a simple remark.

Lemma (1.32). *Suppose that the Markov chain associated with a network (Γ, r) is transient. Let f be a real function on the vertex set V. If $G(|f|)(x) < \infty$ for all x, then $(1 - P)(G(f)) = f$.*

Proof. By the Fubini–Tonelli theorem $P(G(f)) = (PG)(f)$, where PG represents the matrix product. Clearly $(PG)(f) = G(f) - f$. Thus we have

$$(1 - P)(G(f)) = G(f) - P(G(f)) = G(f) - (PG)(f) = f.$$

□

For instance, suppose that the associated Markov chain is transient and that the 0–chain $\imath$ in (1.16) is finite i.e., f is finitely supported. Let $u(x) = G(f)(x)$. Then u is well defined and satisfies (1.27).

DEFINITION. Let (Γ, r) be a network and let P be the associated random walk operator. Let $V' \subseteq V$, $V' \neq \emptyset$. A function u on V is said to be harmonic on V' if $u(x) = P(u)(x)$ for all $x \in V'$. If $V' = V$ we will also say that u is harmonic on (Γ, r).

If $u(x) \geq P(u)(x)$ $(u(x) \leq P(u)(x))$ for all $x \in V'$ we say that u is superharmonic (subharmonic) on V'. If $V' = V$ we will also say that u is superharmonic (subharmonic) on (Γ, r).

Note that if (Γ, r) is transient, then by (1.29) $G(\cdot, y)$ is superharmonic on V and harmonic on $V' = V \setminus \{y\}$, for every $y \in V$.

The following theorem is a well known result in Markov chains theory.

Theorem (1.33). *Let u be a nonnegative function on the vertex set of a recurrent network. Then, u is superharmonic if and only if it is constant.*

Proof. For any fixed $s \in V$ let the initial probability distribution p be assigned so that $p(s) = 1$, $p(x) = 0$ otherwise. Denote by $X_0, X_1, X_2, \ldots$, V-valued random variables such that X_0 has distribution p and

$$p(x, y) = \Pr(X_{n+1} = y | X_n = x)$$
$$= \Pr(X_{n+1} = y | X_n = x, X_{n-1} = t, \ldots, X_0 = s)$$

Let Σ_n be the Borel subalgebra generated by $X_0, X_1, \ldots, X_n$. It is immediately verified that $(u(X_n), \Sigma_n)$ is a supermartingale. Therefore $u(X_n)$ converges almost surely to a finite limit.

Suppose that there are x and y such that $u(x) \neq u(y)$. Since X_n visits x and y infinitely often with probability 1, $u(X_n)$ is equal to $u(x)$ and to $u(y)$ infinitely often with probability 1. Therefore $u(X_n)$ cannot converge almost surely. $\square$

EXAMPLES (1.34). (1) The functions $u_j(x_1, \ldots, x_n) = x_j$, $1 \leq j \leq n$ are easily seen to be harmonic on the n-dimensional grid of Example (1.26) (1).

In the case $n = 2$ all the harmonic polynomials were determined in Stöhr (1950).

(2) Let T be the homogeneous tree of degree 3 (see Figure 1.7), and let all the resistances be 1. Any two neighboring vertices a and b disconnect the tree in two subtrees T_+ and T_-. Denote by V_+ and V_- the respective vertices. Then, $V = V_+ \cup V_-$, $a \in V_+$ and $b \in V_-$. Define a function u on V by

$$u(x) = \begin{cases} \frac{1}{3 \cdot 2^n} - 1 & \text{if } x \in V_+ \text{ and } d(x, a) = n \\ -\frac{1}{3 \cdot 2^n} & \text{if } x \in V_- \text{ and } d(x, b) = n. \end{cases}$$

One checks immediately that u is harmonic. It is also clear that a similar construction can be carried out for every tree of any degree of homogeneity greater than 2.

Now we prove the discrete version of the maximum principle for harmonic functions.

DEFINITION. Let L be a subset of vertices of Γ. We say that a vertex $x \in L$ is a (combinatorial) boundary point of L if there are neighbors of x which are not in L. Otherwise x is called a (combinatorial) interior point .

Theorem (1.35). *Let (Γ, r) be network. Let $V' \subseteq V$ be a connected subset of vertices having connected interior. Suppose that for every boundary point $x \in V'$ there exists at least an interior point $y \in V'$ such that $x \sim y$.*

Let u be a function defined on V and superharmonic on the set of all interior points of V'. If u takes its minimum at an interior point of V', then u is constant on V'. In particular, if u is harmonic in the interior of V' and nonconstant on V', then u attains its maximum and minimum on the boundary of V'.

Proof. Suppose that there exists an interior point $z \in V'$ such that $m = u(z)$ is the minimum of u on V'. By superharmonicity, $u(x) = m$ for all neighbours x of z. By connectedness and by the assumption on the boundary, for every vertex $x \in V'$ there is a sequence $x_0 = z, \dots, x_n = x$ of vertices in V' such that $x_j \sim x_{j+1}$, $j = 0, \dots, n-1$, and x_j is an interior point, for $0 \leq j < n$. Then $m = u(x_0) = u(x_1) = \cdots = u(x_n)$. $\square$

Corollary (1.36). *If (Γ, r) is a finite network, then every superharmonic function on (Γ, r) is constant.*

An analogous result holds in the case of infinite networks if we assume that $\lim u(x) = 0$ as $x \to \infty$ i.e., for every $\epsilon > 0$ there is a finite subset $U \subset V$ such that $|u(x)| < \epsilon$ for all $x \notin U$.

Corollary (1.37). *Suppose that (Γ, r) is infinite. If u is harmonic on (Γ, r) and $u(x) \to 0$ as $x \to \infty$, then $u(x) = 0$ for all $x \in V$.*

Proof. Fix $v \in V$. For every ϵ there is n such that $|u(x)| < \epsilon$ if $d(x, v) \geq n$. Then, by Theorem (1.35), $|u(x)| < \epsilon$ if $d(x, v) \leq n$. $\square$

The following Theorem is known as the Riesz decomposition theorem.

Theorem (1.38). *Suppose that (Γ, r) is a transient network and let u be a non-negative superharmonic function on V. Then, there exist unique functions f and v such that v is harmonic and*

$$(1.39) \qquad\qquad u = G(f) + v.$$

In this decomposition v is the greatest harmonic minorant of u, $f = (1 - P)(u)$ and v and f are both non-negative.

Proof. Since u is non–negative and superharmonic we have

$$u(x) \geq P^n(u)(x) \geq P^{n+1}(u)(x) \geq 0$$

for all $x \in V$ and n. Let $v(x) = \lim_{n \to \infty} P^n(u)(x)$. By the dominated convergence theorem v is harmonic and non–negative. Set $f = u - P(u)$. Then f is non–negative and $u = P^{n+1}(u) + \sum_{j=0}^{n} P^j(f)$. Letting n tend to infinity we have (1.39), by the monotone converge theorem.

If $u = v' + G(f')$, with v' and f' satisfying the same conditions as v and f, we have $(1 - P)(v + G(f)) = (1 - P)(v' + G(f'))$. By Lemma (1.32) $f = f'$. It follows that $v = v'$.

Finally, let h be harmonic and suppose $h(x) \leq u(x)$ for all x. Then $h = P^j(h) \leq P^j(u)$ for all j and, passing to the limit as $j \to \infty$, $h \leq v$. $\square$

REMARK. There is a large body of literature concerning random walks on graphs and discrete harmonic functions. We refer to the survey paper by Woess (1992).

CHAPTER II

FINITE NETWORKS

§1. Existence and uniqueness of currents

In this chapter (Γ, r) will always be a finite network. Under this assumption, Theorem (1.24) and Corollary (1.36) imply that the homogeneous Kirhhoff's equations

$$\partial I = 0$$
$$\langle \mathcal{R}(I), Z \rangle = 0 \quad \text{for every cycle } Z.$$

have only the trivial solution $I = 0$. Thus equations (1.14)–(1.15) have at most one solution. We will prove below that a solution actually exists if and only if $\partial \jmath = 0$. Let us start with an obvious property which has been already remarked in Chapter I.

Lemma (2.1). *Let* $I = \{i(x, y)\}$ *be a 1–chain on the finite graph* Γ. *Then*

$$\sum_{x \in V} \sum_{y \in V(x)} i(x, y) = 0.$$

Proof. Obvious since $\partial \partial I = 0$. $\square$

Since every 1–chain I can be viewed as a flow generated by the 0–chain $\imath = -\partial I$, Lemma (2.1) asserts that the flow entering a finite network is equal to the flow leaving the network. For instance let $\partial I = \kappa(\delta_a - \delta_b)$. Then $\sum_{y \sim a} i(a, y) = \sum_{y \sim b} i(b, y)$.

Theorem (2.2). *Suppose that* (Γ, r) *is a finite network. Let* $\jmath = \sum_{x \in V} \jmath(x) \delta_x$ *be a 0–chain such that*

(2.3) $$\partial \jmath = 0.$$

Let F *be any 1–cochain. Then there is a unique current generated by* $\jmath$ *and* F.

Proof. We will give two proofs. The first proof is more elementary, but the second one provides a probabilistic interpretation of electrical quantities; see Doyle and Snell (1984).

1st PROOF. By (2.3) and Proposition (1.13) the 0–chain $\jmath$ is a boundary. Let J be such that $\partial J = \jmath$ and set $E = F + \mathcal{R}(J)$. Hence we transform Kirchhoff's equations (1.14), (1.15) into the form (1.18), (1.19). Thus we look for a cycle I^* such that

$$(2.4) \qquad \langle \mathcal{R}(I^*), Z \rangle = \langle E, Z \rangle$$

for all cycles Z.

Since all the vector spaces involved are finite dimensional, we may identify 1–chains and 1–cochains. Then $\langle \mathcal{R}(\cdot), \cdot \rangle$ defines a symmetric positive definite bilinear form on the vector space of all 1–chains. Therefore $\mathcal{R}^{-1}(E)$ can be uniquely decomposed as the sum of two 1–chains I^* and K such that I^* is a cycle and K is orthogonal to the cycles space. For such an I^* we have

$$\langle E, Z \rangle = \langle R(R^{-1}(E)), Z \rangle = \langle \mathcal{R}(I^*), Z \rangle$$

i.e., (2.4) holds. Hence $I = I^* + J$ is the unique solution of equation (1.14) and (1.15).

2nd PROOF. We transform Kirchhoff's equations into the equivalent form (1.16), (1.17) as explained in Chapter I. Since $\imath = \jmath + \partial E$, by (2.3) we have $\partial \imath = 0$. By Proposition (1.13), $\imath$ is a boundary, so that $\imath$ is a linear combinations of dipoles, i.e. $\imath = \sum_{k=1}^{n} \lambda_k (\delta_{b_k} - \delta_{a_k})$. Hence, by linearity, it suffices to prove existence in the case where $\imath$ is a dipole.

We proceed as in Doyle and Snell (1984). Let $\imath = \kappa(\delta_b - \delta_a)$. We have to solve equation (1.23) with $f(a) = \kappa c(a)^{-1}$, $f(b) = -\kappa c(b)^{-1}$ and $f(x) = 0$ if $x \neq a$ and $x \neq b$. Let $v(x) = H(x, a; b)$ denote the probability of visiting a before b when the process starts at x. Then $v(a) = 1$, $v(b) = 0$ and, by the second equation of (1.28),

$$v(x) - \sum_y p(x,y)v(y) = \begin{cases} 0 & \text{if } x \neq a, b \\ 1 - \sum_y p(a,y)v(y) & \text{if } x = a \\ -\sum_y p(b,y)v(y) & \text{if } x = b. \end{cases}$$

Note that

$$(2.5) \qquad 1 - \sum_y p(a,y)v(y) > 0.$$

Otherwise v would be harmonic on $V \setminus \{b\}$ and would have maximum value at a, against Theorem (1.35). Let u be the function

$$(2.6) \qquad u(x) = \frac{\kappa v(x)}{c(a)(1 - \sum_y p(a,y)v(y))}.$$

Then u is harmonic on $V \setminus \{a,b\}$ and $(1-P)(u)(a) = f(a)$. We have also $(1 - P)(u)(b) = f(b)$ since, by Lemma (2.1) applied to the 1–chain I such that $i(x,y) = c(x,y)(v(x) - v(y))$, we have

$$
\begin{aligned}
0 &= \sum_{x \in V} \sum_{y \in V(x)} i(x,y) \\
&= c(a)\big(1 - \sum_y p(a,y)v(y)\big) - c(b)\big(\sum_y p(b,y)v(y)\big).
\end{aligned}
$$

Therefore

$$
c(a)\big(1 - \sum_y p(a,y)v(y)\big) = c(b)\big(\sum_y p(b,y)v(y)\big),
$$

and this ends the proof. □

REMARK. Observe that by Theorem (1.13) condition (2.3) is also necessary for the existence of a solution.

In the case of a current generated by a dipole, the potential is expressed by (2.6). The expression for u in the general case can be obtained by taking a linear combination of potentials of currents generated by dipoles. Another expression of the potential of a current in a finite network is provided by the following theorem.

Theorem (2.7). *Let (Γ, r) be a finite network and let ι be a 0–chain such that $\partial \iota = 0$. Let $f(x) = -c(x)^{-1}\iota(x)$ and let b be any vertex in V. The potential u such that $u(b) = 0$ of the current generated by ι is the unique solution of the "boundary problem"*

(2.8)
$$
\begin{aligned}
(1-P)(u)(x) &= f(x) \qquad \text{if } x \neq b, \\
u(b) &= 0.
\end{aligned}
$$

The solution u has the form

(2.9)
$$
u(x) = \sum_{y \in V} G(x,y;b)f(y).
$$

Proof. The solution of problem (2.8) is unique since, by Theorem (1.35), any solution of the homogeneous problem must take its maximum and minimum at b. Let us show that the potential of the current generated by ι, where the additive constant is chosen in such a way that $u(b) = 0$, satisfies (2.8).

Let, as in Chapter I, $\{p(x,y;b)\}$ denote the Markov chain where b has been made absorbing. By definition, $G(b,y;b) = 0$ for every y and $G(x,b;b) = 0$ for every x.

Moreover, if $x \neq b$, then $p_b(x,y) = p(x,y)$. By (1.30) applied to the Markov chain $\{p(x,y;b)\}$ we have, for every $x \neq b$,

$$
\begin{aligned}
(1 - P)(u)(x) &= u(x) - \sum_y p(x,y) \sum_{z \in V} G(y,z;b) f(z) \\
&= u(x) - \sum_{z \neq b} f(z) \sum_{y \neq b} p(x,y) G(y,z;b) \\
&= u(x) + f(x) - \sum_z G(x,z;b) f(z) = f(x).
\end{aligned}
$$

$\square$

§2. The effective resistance

The function u in (2.6) is the potential of the current generated by the dipole $\kappa(\delta_b - \delta_a)$. In particular one obtains

$$
u(a) - u(b) = u(a) = \frac{\kappa}{c(a)\left(1 - \sum_y p(a,y)v(y)\right)}.
$$

If one chooses $\kappa = c(a)\left(1 - \sum_y p(a,y)v(y)\right)$, then $v = u$ and the probability v coincides with the potential of the current flowing in the network when a unit voltage generator is applied between a and b.

DEFINITION. The number

$$
(2.10) \qquad R = R(a,b) = \frac{u(a) - u(b)}{\kappa} = \frac{1}{c(a)\left(1 - \sum_y p(a,y)v(y)\right)}.
$$

is called the effective resistance of the network between a and b. Its inverse $C = R^{-1}$ is called the effective conductance between a and b. A method for computing the effective resistance between two nodes in terms of spanning tree is given in Bollobàs (1979); see also Shapiro (1987).

Effective resistances between neighboring nodes satisfy the celebrated Foster's averaging formula for finite graphs (see Foster (1949), Flanders (1974), Thomassen (1990)). We present below the simple proof due to Tetali (1991). See also Tetali (1994).

Theorem (2.11). *Let (Γ, r) be a finite network. For all $x \sim y$ let $R(x,y)$ denote the effective resistance between x and y. Let n denote the number of vertices of Γ. Then*

$$
\sum_{x \sim y} \frac{R(x,y)}{r(x,y)} = 2(n - 1).
$$

Proof. Fix any vertex v. Then, setting $f = (c(x)^{-1}\delta_x - c(y)^{-1}\delta_y)$ we have by Theorem (2.7)

$$R(x,y) = G(f;v)(x) - G(f;v)(y)$$
$$= c(x)^{-1}(G(x,x;v) - G(y,x;v)) + c(y)^{-1}(G(y,y;v) - G(x,y;v)).$$

Therefore

$$\frac{1}{2}\sum_{x \sim y}\frac{R(x,y)}{r(x,y)} = \sum_{x \neq v}\sum_{y \sim x}p(x,y)(G(x,x;v) - G(y,x;v))$$
$$= n - 1.$$

$\square$

Let $G(a,x;b)$ be as above. We have from the first and the second equation of (1.28)

$$1 - \sum_{y}p(a,y)v(y) = G(a,a;b)^{-1}.$$

The number $1 - \sum_y p(a,y)v(y)$ is called the escape probability $p = p_{ab}$ of the process (see Doyle and Snell (1984)). By (2.10)

$$p_{ab} = c(a)R(a,b)^{-1}.$$

REMARK. Let u be the the potential of the current $I = du$ generated by the dipole $\imath = \delta_b - \delta_a$. By (2.9) $u(x) = c(a)^{-1}G(x,a;b)$. The current flowing in $[x,y]$ is

$$i(x,y) = c(x,y)(u(x) - u(y))$$
$$= c(x)p(x,y)c(a)^{-1}G(x,a;b) - c(y)p(y,x)c(a)^{-1}G(y,a;b)$$
$$= p(x,y)G(a,x;b) - p(y,x)G(a,y;b).$$

Therefore $i(x,y)$ is the net expected number of times a walker starting at a will move along the edge connecting x and y before reaching b; See Doyle and Snell (1984); see also Tetali (1991) and Konsowa (1992).

We end this section with an interesting formula for the effective resistance in the case of a graph obtained by shorting the terminal vertices of a tree T.

Let $T = (V,Y)$ be a finite tree. Suppose that there exists a vertex x_0 with the property that there is $n > 1$ such that $\max_{x \in V}d(x,x_0) = n$, and $\deg(x) = 1$ if and only if $d(x,x_0) = n$. A vertex x such that $\deg(x) = 1$ is called a terminal vertex of the tree (see Figure 2.1). We introduce the following notation: $x_0 \succ y$ for all $y \in V$. If $x \neq x_0$ and $y \neq x_0$, then we write $x \succ y$ whenever x is a vertex of the

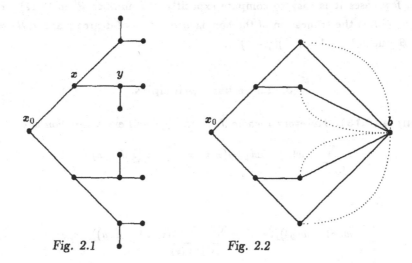

Fig. 2.1   Fig. 2.2

unique path joining x_0 and y. For every $x \in V$ we denote by $U(x)$ the set of all vertices y such that $x \succ y$ and $x \sim y$.

Let (T', r') be the network obtained by shorting together all the terminal vertices of T. See for instance the graph in Figure 2.2. obtained by shorting together all the terminal points of the tree in Figure 2.1. Every dotted edge is identified with the solid edge having the same endpoints.

Denote by $y_1, \ldots, y_k$ the neighbors of x_0. For every $j = 1, \ldots, k$, let Γ_j be the induced subgraph of T' whose vertex set contains only b, y_j and all the vertices $y \in V$ such that $y_j \succ y$. Let R_j denote the effective resistance of Γ_j between y_j and b.

By (2.10) the total resistance R' between x_0 and b of T' is

$$R' = \Big(\frac{1}{r(x_0, y_1) + R_1} + \frac{1}{r(x_0, y_2) + R_2} + \cdots + \frac{1}{r(x_0, y_k) + R_k} \Big)^{-1}.$$

A repeated application of this formula gives
(2.12)

$$R' = \cfrac{1}{\displaystyle\sum_{x_1 \in U(x_0)} r(x_0, x_1) + \cfrac{1}{\displaystyle\sum_{x_2 \in U(x_1)} r(x_1, x_2) + \cfrac{1}{\ddots + \cfrac{1}{\displaystyle\sum_{x_n \in U(x_{n-1})} r(x_{n-1}, x_n)}}}}$$

In a few cases it is easy to compute explicitly the number R' in (2.12). For instance, if T is the truncation of the homogeneous tree of degree q and $r(B) = 1$ for all B, then $R' = (1 - q^{-n})(q - 1)^{-1}$.

§3. Three basic principles

Lemma (2.13). *For every 1–chain $J = \{j(x,y)\}$ and every function w on V*

$$\sum_{[x,y] \in Y} (w(x) - w(y))j(x,y) = 2 \sum_x w(x)\partial J(x).$$

Proof. We have

$$\sum_{x \sim y} (w(x) - w(y))j(x,y) = \sum_{x \in V} \sum_{y \in V(x)} (w(x) - w(y))j(x,y)$$

$$= 2 \sum_{x \in V} w(x) \sum_{y \in V(x)} j(x,y)$$

$$= 2 \sum_{x \in V} w(x)\partial J(x).$$

$\square$

Note that if J is the flow generated by the dipole $\kappa(\delta_b - \delta_a)$, then Lemma (2.13) implies that $\sum_{[x,y] \in Y}(w(x) - w(y))j(x,y) = 2\kappa(w(a) - w(b))$.

DEFINITION. Let I be a 1–chain. The energy (or total power dissipated) of I is defined as

$$\mathcal{W}(I) = \frac{1}{2} \sum_{[x,y] \in Y} r(x,y)i^2(x,y).$$

We deduce the law of conservation of energy in the network from Lemma (2.13).

Corollary (2.14). *Let I be the current generated by a dipole $\kappa(\delta_b - \delta_a)$. Let R be the resistance between a and b and let u be the potential of I. Then*

$$\kappa(u(a) - u(b)) = \kappa^2 R = \mathcal{W}(I).$$

Th following Theorem is known as Thomson's principle.

Theorem (2.15). *Let I be the current generated by a 0–chain $\imath$. Then, for every flow J generated by $\imath$, $\mathcal{W}(I) \leq \mathcal{W}(J)$ and equality holds if and only if $I = J$.*

Proof. Let u be the potential of I and set $D = J - I$. By Lemma (2.13)

$$\sum_{[x,y] \in Y} r(x,y)i(x,y)d(x,y) = \sum_{[x,y] \in Y} (u(x) - u(y))d(x,y) = 0$$

Hence,

$$W(J) = \frac{1}{2} \sum_{[x,y] \in Y} r(x,y)i^2(x,y) + \sum_{[x,y] \in Y} r(x,y)i(x,y)d(x,y) +$$

$$+ \frac{1}{2} \sum_{[x,y] \in Y} r(x,y)d^2(x,y) = W(I) + W(D),$$

thus proving the theorem. $\square$

Set, for all real functions w on V,

$$D(w) = \frac{1}{2} \sum_{[x,y] \in Y} c(x,y)(w(x) - w(y))^2$$

$$= \frac{1}{2} \sum_{x \in V} \sum_{y \in V(x)} c(x,y)(w(x) - w(y))^2.$$

Arguing as in the proof of Thomson's principle we obtain the so called Dirichlet principle (in the dipole case).

Theorem (2.16). *Let R be the effective resistance between a and b. Let u denote the potential such that $u(b) = 0$ of the current generated by the dipole $R^{-1}(\delta_b - \delta_a)$. Then*

$$R^{-1} = D(u) = \inf\{D(w) : w(a) = 1, \; w(b) = 0\}.$$

Proof. First of all note that, by the definition of R, $u(a) = 1$. For every function w on V such that $w(a) = 1$, $w(b) = 0$, set $w = u + d$. Then $d(a) = d(b) = 0$. Moreover

$$\frac{1}{2} \sum_{[x,y] \in Y} c(x,y)(u(y) - u(x))(d(y) - d(x)) =$$

$$= \sum_{x \in V} d(x) \sum_{y \in V(x)} c(x,y)(u(x) - u(y))$$

$$= d(a)c(a)(1 - P)(u)(a) + d(b)c(b)(1 - P)(u)(b) = 0.$$

Hence $D(w) = D(u) + D(d) \geq D(u)$. Suppose that $I = du$ is the current generated by the dipole $R^{-1}(\delta_b - \delta_a)$. Then $D(u) = W(I)$. By Corollary (2.14) $R^{-1} = W(I) = D(u)$. $\square$

The following Theorem is known as Rayleigh's monotonicity principle. The formulation given below is somewhat more general than the one usually found in the literature.

Theorem (2.17). *Let (Γ, r) and (Γ', r') be finite networks, $\Gamma = (V, Y)$, $\Gamma' = (V', Y')$. Suppose that $\phi : V \mapsto V'$ induces a network morphism. Let $a, b \in V$ be such that $\phi(a) \neq \phi(b)$. Then, denoting by R the effective resistance between a, b and by R' the effective resistance between $\phi(a)$ and $\phi(b)$, we have*

$$(2.18) \qquad\qquad\qquad R' \leq R.$$

Proof. Let $I' = \{i'(B')\}$ denote the current generated in (Γ', r') by the dipole $\delta_{\phi(b)} - \delta_{\phi(a)}$ and let $I = \{i(B)\}$ denote the current generated in (Γ, r) by $\delta_b - \delta_a$. Let $\Phi(Y) \subseteq Y'$ denote the image of Y. We define a 1–chain $J = \{j(B')\}$ on (Γ', r') in the following way

$$(2.19) \qquad j(B') = \begin{cases} \sum_{\{B \in Y : \Phi(B) = B'\}} i(B) & \text{if } B' \in \Phi(Y) \\ 0 & \text{otherwise.} \end{cases}$$

Note that J is actually a 1–chain, since (2.19) implies $j(x', y') = -j(y', x')$ for all $x' \sim y'$ in V'.

It is easy to check that J is a flow generated by $\phi(b) - \phi(a)$. In fact, let $(\partial J)(x')$ and $(\partial I)(x)$ denote the value of the boundary of J at $x' \in V'$ and of I at $x \in V$ respectively. If $x' \notin \phi(V)$ then $J(x', y') = 0$ for any $y' \sim x'$. Hence $(\partial J)(x') = 0$. If $x' \in \phi(V)$, then

$$
\begin{aligned}
\partial J(x') &= \sum_{y' \sim x'} j(x', y') \\
(2.20) \qquad\qquad &= \sum_{\{x : \phi(x) = x'\}} \sum_{y \sim x} i(x, y) \\
&= \sum_{\{x : \phi(x) = x'\}} \partial I(x).
\end{aligned}
$$

By (2.20) we see that $\partial J = \delta_{\phi(a)} - \delta_{\phi(b)}$. Now we have by Thomson's principle

$$(2.21) \qquad R' = \mathcal{W}(I') \leq \mathcal{W}(J) = \frac{1}{2} \sum_{B' \in \Phi(Y)} j(B')^2 r'(B').$$

By Cauchy–Schwartz inequality we have

$$j(B')^2 \leq \sum_{\{B \in Y : \Phi(B) = B'\}} i(B)^2 r(B) \sum_{\{B \in Y : \Phi(B) = B'\}} \frac{1}{r(B)}.$$

Thus we get from (2.18) and (2.21)

$$R' \leq \frac{1}{2} \sum_{B \in Y} i^2(B) r(B) = \mathcal{W}(I) = R.$$

$\square$

Rayleigh's principle has the following immediate consequences.

Corollary (2.22). *Decreasing the resistance of one or more edges of a finite network decreases (or does not affect) the effective resistance between two nodes.*

Corollary (2.23). *Shorting together a subset of nodes of a finite network decreases (or does not affect) the effective resistance between two nodes.*

Corollary (2.24). *Cutting one or more edges of a finite network increases (or does not affect) the effective resistance between two nodes.*

CHAPTER III

CURRENTS AND POTENTIALS WITH FINITE ENERGY

§1. Finite energy and Dirichlet sums

In this chapter we return to infinite networks. As already mentioned, an infinite network admits in general nontrivial harmonic functions (see e.g. Examples (1.34)), so that equations (1.14)–(1.15) admit in general more than a solution. In these notes we will restrict the class of admissible solutions of Kirchhoff's equations to the class of currents with finite energy. Even though this restriction is not sufficient to guarantee the uniqueness of a solution, yet this approach allows us to use the functional analytic tools and seems to be the main road in infinite networks theory. For a study of the whole class of different voltage–current regimes that a network can have, we refer to Zemanian (1976)(b) and (1978).

DEFINITION. Let (Γ, r) be a network and let I be a 1–chain on Γ. The energy $\mathcal{W}(I)$ of I is defined as

$$(3.1) \qquad \mathcal{W}(I) = \frac{1}{2} \sum_{B \in Y} r(B) i(B)^2.$$

The norm

$$\|I\|_1 = \mathcal{W}(I)^{1/2}$$

defines a Hilbert space structure $\mathbf{H}_1$ on the linear space of all 1–chains with finite energy.

A 1–cochain $J = \sum_{B \in X} j(B) \chi_B$ can be continuously extended to the whole of $\mathbf{H}_1$ if and only if $\sum_{B \in X} r(B) j(B)^2 < \infty$. Henceforth we identify these 1–cochains with elements of $\mathbf{H}_1$.

Let $(\cdot, \cdot)$ denote the inner product in $\mathbf{H}_1$. We have

$$\langle \mathcal{R}(I), Z \rangle = (I, Z).$$

We let $\mathbf{H}_0$ denote the Hilbert space of all 0–chains $\jmath = \sum_{x \in V} j(x) \delta_x$ such that

$$(3.2) \qquad \|\jmath\|_0^2 = \sum_{x \in V} c(x)^{-1} j(x)^2 < \infty.$$

The energy of $\jmath$ is the sum in (3.2). As before, the space of all 0–cochains which are continuously extendable to $\mathbf{H}_0$ is identified with $\mathbf{H}_0$ itself.

From now on we will consider only Kirchhoff's equations (1.14), (1.15) in which the energy sources have finite energy, and we will look only for solutions $I \in \mathbf{H}_1$.

The following lemma describes the properties of the boundary operator on $\mathbf{H}_1$ which are relevant in the following.

Lemma (3.3). *The space $\mathbf{H}_1$ is contained in $\mathbf{C}$. For every $x \in V$ the linear functional which assigns to $I \in \mathbf{H}_1$ the value $\partial I(x)$ is continuous on $\mathbf{H}_1$. Furthermore, the boundary operator ∂ maps continuously $\mathbf{H}_1$ into $\mathbf{H}_0$.*

Proof. For every $I \in \mathbf{H}_1$ and $x \in V$ we have

$$\sum_{y \in V(x)} |i(x,y)| \le \Big(\sum_{y \in V(x)} c(x,y) \Big)^{1/2} \Big(\sum_{y \in V(x)} r(x,y) i(x,y)^2 \Big)^{1/2}$$
$$\le (c(x)\mathcal{W}(I))^{1/2}.$$

Therefore $\mathbf{H}_1$ is contained in $\mathbf{C}$.

For every $x \in V$ let S^x be the 1–cochain such that $S^x(B) = r(B)^{-1}$ if $B = [x,y]$ is not a self–loop, $S^x(B) = -r(B)^{-1}$ if $B = [y,x]$ is not a self–loop and $S^x(B) = 0$ if x is not an endpoint of B or B is a self–loop. Since

$$\mathcal{W}(S^x) \le \sum_{y \in V(x)} r(x,y) \frac{1}{r(x,y)^2} = c(x) < \infty,$$

we obtain that S^x belongs to $\mathbf{H}_1$. Moreover $\partial I(x) = \sum_{y \in V(x)} i(x,y) = (S^x, I)$, so that the functional $I \mapsto \partial(I)(x)$ is continuous for all x.

Finally,

$$\|\partial I\|_0^2 = \sum_{x \in V} c(x)^{-1} \Big(\sum_{y \in V(x)} i(x,y) \Big)^2$$
$$\le \sum_{x \in V} c(x)^{-1} \Big(\sum_{y \in V(x)} c(x,y) \Big) \Big(\sum_{y \in V(x)} r(x,y) i^2(x,y) \Big)$$
$$= 2\|I\|_1^2.$$

Thus ∂ is continuous from $\mathbf{H}_1$ to $\mathbf{H}_0$. $\square$

Now we write Kirchhoff's equations in the case of finite energy. Suppose that $\jmath \in \mathbf{H}_0$ and $E = \mathcal{R}^{-1}(F) \in \mathbf{H}_1$. A 1–chain $I \in \mathbf{H}_1$ is a current generated by $\jmath$ and F if and only if I satisfies the equations

(3.4) $\qquad\qquad \partial I + \jmath = 0$

(3.5) $\qquad\qquad (I - E, Z) = 0 \qquad$ for all finite cycles Z.

Setting $\imath = \jmath + \partial E$ we have that $\imath \in \mathbf{H}_0$ by Lemma (3.3). Denoting by I again the new unknown $I - E$ equations (3.4), (3.5) become

(3.6) $\partial I + \imath = 0$

(3.7) $(I, Z) = 0$ for all finite cycles Z.

In order that equations (3.4), (3.5) admit a solution with finite energy it is necessary that $\jmath = \partial J$ for some $J \in \mathbf{H}_1$. Therefore we are lead to the following definition.

DEFINITION. We denote by $\partial \mathbf{H}_1$ the subspace of $\mathbf{H}_0$ consisting of all 0–chain which are boundaries of elements in $\mathbf{H}_1$.

Finally, we can write the third form of Kirchhoff's equations. Suppose that $\jmath \in \partial \mathbf{H}_1$. If $\jmath = \partial J$, $J \in \mathbf{H}_1$. We set $E = \mathcal{R}^{-1}(F) + J$ and call I again the new unknown $I + J$. Thus obtain the third form of Kirchhoff's equations

(3.8) $\partial I = 0$

(3.9) $(I - E, Z) = 0$ for all finite cycles Z.

Suppose now that u is a function on V such that $du = I$ for some $I \in \mathbf{H}_1$. Then, u satisfies (1.22) and, by (1.20),

(3.10) $$D(u) = \frac{1}{2} \sum_{[x,y] \in Y} c(x,y)|u(x) - u(y)|^2 < \infty.$$

Conversely, if u satisfies (3.10), then u satisfies also (1.22), since $I = du$ must belong to $\mathbf{H}_1$.

DEFINITION. For every (real–valued) function u on V the quantity $D(u)$ in (3.10) is called the Dirichlet sum of u. We will also say that a function satisfying (3.10) is Dirichlet finite.

Note that $D(u)$ is the discrete analogue of the energy integral on a Riemann manifold.

Arguing as in §4 of Chapter I, we obtain Poisson's equation from (3.6), (3.7)

(3.11) $$u(x) - \sum_{y \in V} p(x,y)u(y) = f(x)$$

where $f(x) = -c(x)^{-1}\imath(x)$. Clearly $\imath \in \mathbf{H}_0$ if and only if $\sum_{x \in V} c(x)f(x)^2 < \infty$.

DEFINITION. We denote by ℓ^2 the Hilbert space of all real valued functions f on V such that

$$\|f\|_2^2 = \sum_{x \in V} c(x)|f(x)|^2 < \infty.$$

Clearly the application mapping $\imath$ to f, where $f(x) = -c(x)^{-1}\imath(x)$, is an isometry of $\mathbf{H}_0$ onto ℓ^2.

Lemma (3.12). *The operator P is continuous and hermitian on ℓ^2 with norm not greater than 1.*

Proof. We have for all $f \in \ell^2$

$$\sum_x c(x)\Big(\sum_y p(x,y)|f(y)|\Big)^2 = \sum_{y,z} |f(y)f(z)|w(y,z)$$

where $w(y,z) = \sum_x c(x)p(x,y)p(x,z)$. Also

$$\sum_{y,z} |f(y)f(z)|w(y,z) \le \Big(\sum_{y,z} |f(y)|^2 w(y,z)\Big)^{1/2}\Big(\sum_{y,z} |f(z)|^2 w(y,z)\Big)^{1/2}$$

$$= \sum_{y,z} |f(y)|^2 w(y,z)$$

since $w(y,z) = w(z,y)$. Since

$$\sum_z w(y,z) = \sum_z p(x,y)c(x)\sum_z p(x,z) = \sum_x p(x,y)c(x)$$

$$= \sum_z p(y,x)c(y) = c(y)$$

we have $\sum_{y,z} |f(y)|^2 w(y,z) = \sum_y c(y)|f(y)|^2$.

Hence the norm of the operator P is not greater than 1. It is now routine to check that, for all $f, g \in \ell^2$, $\sum_x c(x)P(f)(x)g(x) = \sum_x c(x)f(x)P(g)(x)$. $\square$

By (3.10) a function on V is the potential of a current in $\mathbf{H}_1$ if and only if it has finite Dirichlet sum. We can state the analogue of Theorem (1.24).

Theorem (3.13). *For every solution $I \in \mathbf{H}_1$ of (3.6), (3.7), with $\imath \in \mathbf{H}_0$, there is a function u, unique up to an additive constant, such that $D(u) < \infty$, u satisfies (3.11) and $du = I$. Conversely, for every solution u, such that $D(u) < \infty$, of Poisson's equation (3.11), where $f \in \ell^2$, there is a unique solution $I \in \mathbf{H}_1$ of (3.6), (3.7) such that $du = I$. Kirchhoff's equations have at most one solution in $\mathbf{H}_1$ if and only if the unique harmonic functions with finite Dirichlet sum are the constants.*

DEFINITION. Let ℓ_0 denote the linear space of all real–valued finitely supported functions on V and let o be a fixed reference vertex.

We denote by $\mathbf{D}$ the space of all real–valued functions on u on V, satisfying (3.10), endowed with the norm

$$\|u\| = \big(c(o)|u(o)|^2 + D(u)\big)^{1/2}.$$

We denote by $\mathbf{D}_0$ the completion of ℓ_0 in the above norm.

Lemma (3.14). *For every $x \in V$ there is a constant $K_x > 0$ such that for every $u \in \mathbf{D}$*

$$|u(x)| \le K_x \|u\|.$$

Proof. Let $n = d(x, o)$. For any path with vertices $x_0 = o \sim x_1 \sim \cdots \sim x_n = x$ connecting o and x we have

$$|u(x)| \le \sum_{k=1}^{n} |u(x_k) - u(x_{k-1})| + |u(o)| \le \left(c(o)^{-1} + \sum_{k=1}^{n} r(x_{k-1}, x_k) \right)^{1/2} \|u\|.$$

$\square$

Theorem (3.15). *$\mathbf{D}$ is a Hilbert space and $\mathbf{D}_0$ is a closed subspace of $\mathbf{D}$. If a sequence of functions $u_n \in \mathbf{D}$ converges weakly to $u \in \mathbf{D}$, then $\lim_{n \to \infty} u_n(x) = u(x)$ for all $x \in V$.*

Proof. The proof that $\mathbf{D}$ is complete is standard. Suppose that $u_n \in \mathbf{D}$ is a Cauchy sequence. By Lemma (3.14) $u_n(x)$ is a Cauchy sequence for all $x \in V$. Let $u(x) = \lim_{n \to \infty} u_n(x)$. Choose $\epsilon > 0$. For large values of n and m, $\|u_n - u_m\| < \epsilon$. By Fatou's lemma $\|u - u_m\|^2 \le \liminf_{n \to \infty} \|u_n - u_m\|^2 < \epsilon$, so that $u \in \mathbf{D}$ and $\|u - u_m\| \to 0$. In particular $\mathbf{D}_0$ is a closed subspace. Finally, the last assertion follows from the fact that evaluation at a point is a continuous functional by Lemma (3.14) $\square$

REMARK. Clearly $\mathbf{D}$ and $\mathbf{D}_0$ do not depend on the reference vertex o. Suppose that $o' \in V$ and let $\mathbf{D}'$ be defined as $\mathbf{D}$, with o' in place of o. Then, by Lemma (3.14), $\mathbf{D}$ and $\mathbf{D}'$ are isomorphic Hilbert spaces.

DEFINITION. The space $\mathbf{D}$ is called the Dirichlet space associated with the network Γ. It is immediate to see that the inner product in $\mathbf{D}$ is a Dirichlet form in the sense of Fukushima (1980).

§2. Existence of currents with finite energy

In this section we will prove two existence theorems for currents with finite energy. The first one is due to H. Flanders (see Flanders (1971)), while the second one is an extension of a Theorem by Zemanian (see Zemanian (1976)(a)). These theorems state also that the current is unique in $\mathbf{H}_1$ provided that it satisfies an extra condition: in Flander's theorem uniqueness holds if we require that the current is the limit in $\mathbf{H}_1$ of finite currents. In Zemanian's setup it is required that the current satisfies a stronger form of Kirchhoff's loop law.

Theorem (3.16). *Let (Γ, r) be a network and let E be a finite 1–chain. Let $\mathbf{Z}$ denote the space of all finite cycles. Then, the closure in $\mathbf{H}_1$ of $\mathbf{Z}$ contains a unique solution of Kirchhoff's equations (3.8), (3.9).*

Proof. The proof is a direct generalization of the first proof of Theorem (2.2). There is a unique cycle I in the closure of $\mathbf{Z}$ such that $(I, Z) = (E, Z)$ for all finite cycles Z. This is obvious, since E can be uniquely decomposed as $E = I + K$, where $I \in \mathbf{Z}$ and $K \in \mathbf{H}_1$ is orthogonal to $\mathbf{Z}$. □

Corollary (3.17). *Let $\{(\Gamma_n, r_n)\}$ be any exhaustion of (Γ, r) such that the support of E lies in V_1. Denote by I_n the currents generated by E in (Γ_n, r_n) and by I the current dictated by Theorem (3.16). Extend I_n to a 1–chain on Γ by setting $i_n(B) = 0$ if $B \notin Y_n$. Then $\|I - I_n\|_1 \to 0$ as $n \to \infty$.*

Proof. As in the previous proof we let $\mathbf{Z}$ denote the closure in $\mathbf{H}_1$ of the space of all finite cycles. We denote by $\mathbf{Z}_n$ the closed subspace of all cycles whose edges are in Y_n. The first proof of Theorem (2.2) and Theorem (3.16) above show that I_n is the projection of E onto $\mathbf{Z}_n$ and I is the projection of E onto $\mathbf{Z}$.

For all $Z \in \mathbf{Z}_n$ we have $(I_n, Z) = (I, Z)$. For every $\epsilon > 0$ there is n_0 and a cycle J belonging to $\mathbf{Z}_n$ for every $n \geq n_0$ such that $\|J - I\|_1 < \epsilon$. For all $n \geq n_0$ and $Z \in \mathbf{Z}_n$

$$(3.18) \qquad |(I_n - J, Z)| = |(I - J, Z)| \leq \|I - J\|_1 \|Z\|_1 < \epsilon \|Z\|_1.$$

Since $I_n - J$ is in $\mathbf{Z}_n$, (3.18) implies $\|I_n - J\|_1 < \epsilon$. Therefore $\|I_n - I\|_1 < 2\epsilon$ for $n \geq n_0$. □

EXAMPLE (3.19). The homogeneous tree example (see example (1.34)(2)) shows that Kirchhoff's equations in general admit more than one solution in $\mathbf{H}_1$. In fact the harmonic function u described in Example (1.34)(2) belongs to $\mathbf{D}$. Namely, there are 2^k points $x \in V_+$ such that $d(x, a) = k$ and 2^k points $x \in V_-$ such that $d(x, b) = k$. It follows that

$$D(u) = \frac{1}{9} + \frac{1}{9} \sum_{k=1}^{+\infty} 2^k 2^{-2k} + \frac{1}{9} \sum_{k=1}^{+\infty} 2^k 2^{-2k} = 1/3.$$

REMARK. Let E be a finite 1–chain and let I be the current dictated by Flanders' Theorem. The 1–chain $I - E$ satisfies (3.6), (3.7) with $\imath = \partial E$. Conversely, suppose that we want to solve equation (3.6), (3.7) where $\imath$ is the boundary of a finite 1–chain. As already observed, the solutions I' of (3.6), (3.7) and I of (3.8), (3.9) are related by $I' = I - E$, where E is *any* 1–chain such that $\partial E = \imath$. It is immediately seen that $I' = I - E$ does not depend on E provided that E is finite.

DEFINITION. Suppose that $i = \partial E$, where E is a finite 1–chain. Let I be the 1–chain in $\mathbf{H}_1$ satisfying (3.8), (3.9) and lying in the closure of the space of all finite cycles. We will say that the 1–chain $I - E$ is the limit current generated by i. Thus, the limit current satisfies (3.6), (3.7) and, by the above Remark, the definition is unambiguous.

The limit current will be denoted by I_L.

Let (Γ_n, r_n) be an exhaustion of the network (Γ, r) and let $a, b \in V_1$. By Rayleigh's principle (Theorem (2.17)), the effective resistances R_n between a and b of the networks (Γ_n, r_n) form a decreasing sequence.

DEFINITION. The limit $R_L = \lim_{n \to \infty} R_n$ is called the limit resistance of the network between a and b.

Let u_L denote the potential of the limit current. If u_n is the potential of the current I_n in (Γ_n, r_n), then $u_n(x)$ converges to $u_L(x)$ for every $x \in V$. Thus we obtain from (2.14) the law of conservation of energy (for the limit current) in infinite networks.

Corollary (3.20). *Let I_L be the limit current generated by the dipole $\delta_b - \delta_a$, and let $du_L = I_L$ be the corresponding potential. Then*

$$u_L(a) - u_L(b) = R_L = \mathcal{W}(I_L).$$

Let $\mathbf{Z}^*$ denote the space of all cycles, finite or infinite, in $\mathbf{H}_1$. Since the map $Z \mapsto \partial Z(x)$ is continuous for all x, we see that $\mathbf{Z}^*$ is closed. Let $\tilde{\mathbf{Z}}$ be any closed subspace of $\mathbf{Z}^*$ containing all the finite cycles. Then $\tilde{\mathbf{Z}}$ "determines" a unique solution of equation (3.8), (3.9).

Theorem (3.21). *Let $\tilde{\mathbf{Z}}$ be any closed subspace of $\mathbf{Z}^*$ containing all the finite cycles. For every $E \in \mathbf{H}_1$ there exists a unique $\tilde{I} \in \tilde{\mathbf{Z}}$ satisfying the following strengthened form of Kirchhoff's equations (3.8), (3.9)*

$$
\begin{aligned}
\partial \tilde{I} &= 0, \\
(\tilde{I} - E, Z) &= 0 \qquad \text{for all cycles } Z \in \tilde{\mathbf{Z}}.
\end{aligned}
$$
(3.22)

Such a current is the orthogonal projection of E onto the space $\tilde{\mathbf{Z}}$. Furthermore, if $\tilde{\mathbf{Z}}_1 \subseteq \tilde{\mathbf{Z}}_2$, then we have for the corresponding currents $\tilde{I}_1$ and $\tilde{I}_2$

$$\mathcal{W}(\tilde{I}_2 - E) \leq \mathcal{W}(\tilde{I}_1 - E).$$

Proof. The proof is immediate. In fact the solution $\tilde{I}$ is the projection of E onto $\tilde{\mathbf{Z}}$. Note that

$$\mathcal{W}(E) = \mathcal{W}(\tilde{I}) + \mathcal{W}(\tilde{I} - E).$$
(3.23)

If $\tilde{Z}_1 \subseteq \tilde{Z}_2$, then $\mathcal{W}(\tilde{I}_1) \leq \mathcal{W}(\tilde{I}_2)$. By (3.23) the projections on the complementary spaces satisfy the reverse inequality. $\square$

REMARK. If $\tilde{Z} = Z$ is the closure of the space of all finite cycles, then $\tilde{I} - E$ has maximum energy among the currents dictated by the preceding Theorem. If E is finite then $I - E$ coincides with the limit current I_L generated by $\imath = \partial E$. If $\tilde{Z}$ is the whole cycle space Z^*, then the 1–chain $I^* - E$ has minimum energy.

Here we made $I^* = \tilde{I}$. Clearly $I^* - E$ satisfies (3.6), (3.7) with $\imath = \partial E$.

Now, fix $\imath \in \partial \mathbf{H}_1$ and let E be any 1–chain in $\mathbf{H}_1$ such that $\partial E = \imath$. It is immediately seen that if I^* is the solution of (3.22) with $\tilde{Z} = Z^*$, then $I^* - E$ does not depend on E. Hence the following definition is unambiguous.

DEFINITION. Let $\imath = \partial E$ for some $E \in \mathbf{H}_1$ and let Z^* be the subspace of all cycles in $\mathbf{H}_1$. Let I^* be the unique solution of equations (3.22) when $\tilde{Z} = Z^*$. We will say that $I^* - E$ is the minimal current generated by $\imath$. Clearly $I^* - E$ satisfies (3.6), (3.7) with $\imath = \partial E$.

The minimal current will also be denoted by I_M.

§3. The minimal current

In this section we will show that, in analogy with the limit current, the minimal current I_M can be obtained as a limit of currents in finite networks.

We start by describing the relation between 1–chains with finite energy on a network (Γ, r) and 1–chains with finite energy on the network (Γ', r') obtained by shorting together a subset $U \subseteq V$ by means of a morphism $\phi : V \mapsto V$.

DEFINITION. Let (Γ, r) and let $U \subseteq V$ be a finite subset. Suppose that (Γ', r') is obtained by shorting together the vertices in $V \setminus U$ (see §1 in Chapter I). For every 1–chain I with finite energy on Γ we define the restriction J of I to (Γ', r') as the 1–chain on Γ' whose value $j(B')$ on $B' \in Y'$ is given by

$$(3.24) \qquad j(B') = \begin{cases} \sum_{\{B:\Phi(B)=B'\}} i(B) & \text{if } B' \text{ is not a self–loop} \\ 0 & \text{otherwise.} \end{cases}$$

The 1–chain J is well defined and $\mathcal{W}(J) \leq \mathcal{W}(I)$, since

$$\sum_{B' \in Y'} r'(B')|j(B')|^2 \leq$$

$$\sum_{B' \in Y'} r'(B') \sum_{\{B \in Y: \Phi(B)=B'\}} \frac{1}{r(B)} \sum_{\{B \in Y: \Phi(B)=B'\}} r(B)i^2(x,y) = \mathcal{W}(I).$$

Conversely we define the extension of a 1–chain J on Γ' as the 1–chain I on Γ such that

$$i(B) = j(B')r'(B')/r(B) \quad \text{if } \Phi(B) = B' \in Y'.$$

Note that $i(B) = 0$ if B maps to a self-loop. Hence $i(x,y) = 0$ if both x and y belong to $V \setminus U$. The 1-chain I is well defined and $\mathcal{W}(I) = \mathcal{W}(J)$, since

$$\mathcal{W}(I) = \sum_{B \in Y} r(B) i(B)^2$$
$$= \sum_{B' \in Y'} r'(B') j(B')^2 \sum_{\{B: \, \Phi(B) = B'\}} r'(B')/r(B)$$
$$= \mathcal{W}(J).$$

Note also that, by the definition of r', $\partial I(x) = \partial' J(x)$ for all $x \in U$. Here ∂' denotes the boundary operator on Γ'.

Suppose now that (Γ, r) is an infinite network and let $\{(\Gamma_n, r_n)\}$ be an exhaustion of (Γ, r) (as usual we set $\Gamma_n = (V_n, Y_n)$).

For every n we define a new network (Γ'_n, r'_n) by shorting together by means of a morphism $\phi_n : V \mapsto V$ all the vertices in $V \setminus V_n$. Let V'_n and Y'_n denote the vertex and edge set of Γ'_n respectively. Let b_n be the vertex of V'_n which is not a vertex of V_n. Then $V'_n = V_n \bigcup \{b_n\}$ and $Y_n \subset Y'_n$. For all $x \sim y \in V_n$ the conductances $c(x,y)$ are the same in (Γ, r) and (Γ'_n, r'_n). Analogously, if $x \in V_n$, then the total conductances at x are the same in the two networks.

Suppose now that $\imath$ is a 0-chain on Γ. For every n we define a 0-chain $\imath_n$ on Γ'_n in the following way:

$$\imath_n(x) = \begin{cases} \imath(x) & \text{if } x \in V_n \\ -\sum_{y \in V_n} \imath(x) & \text{if } x = b_n. \end{cases}$$

The following theorem was proved independently by Zemanian (1991)(b) and Schlesinger (1992). The shorting method employed in the proof of the theorem is a technique due mainly to Doyle and Snell (1984) and T. Lyons (1983).

Theorem (3.25). *Suppose that $\imath \in \partial \mathbf{H}_1$. Let I'_n be the unique current generated by $\imath_n$ in (Γ'_n, r'_n) and let I_n be its extension to Γ. Then $\mathcal{W}(I'_n)$ is a bounded sequence and I_n converges in $\mathbf{H}_1$ to the minimal current I_M generated by $\imath$ in (Γ, r). Furthermore, $\mathcal{W}(I_M) < \mathcal{W}(J)$ for every other flow generated by $\imath$.*

Proof. Let I^n be the restriction of I_M to Γ'_n and let ∂_n denote the boundary operator on Γ'_n. For every $x \in V_n$ the value of $\partial_n I^n$ at x is equal to the value of ∂I_M at x. Since $\partial I_M + \imath = 0$, by Lemma (2.1) and the definition of $\imath_n$, $(\partial_n I^n)(x) + \imath_n(x) = 0$ for every $x \in V'_n$. Therefore I^n is a flow generated by $\imath_n$.

Let I'_n be the current generated in (Γ', r'_n) by $\imath_n$ and let I_n denote its extension to (Γ, r). By Thomson's principle for finite networks

$$\mathcal{W}(I_n) = \mathcal{W}(I'_n) \leq \mathcal{W}(I^n) \leq \mathcal{W}(I_M).$$

Since $\mathcal{W}(I_n)$ is bounded there exists a subsequence I_{n_k} weakly convergent in $\mathbf{H}_1$ to a 1–chain J on Γ. Then, for every edge $B \in Y$, $I_{n_k}(B)$ converges to $J(B)$. Moreover, by Lemma (3.3), $\partial I_{n_k}(x)$ converges to $\partial J(x)$ for every $x \in V$. If k is sufficiently large (depending on x)

$$\partial I_{n_k}(x) = \partial_{n_k} I'_{n_k}(x) = -\imath(x),$$

so that $\partial J + \imath = 0$. Hence J satisfies the node law.

Given any finite cycle Z on Γ, for large values of k the edges B such that $z(B) \neq 0$ belong to Y_{n_k} and $r(B) = r'_{n_k}(B)$. It follows that the restriction Z^n of Z to Γ'_n is a cycle on Γ'_n and that

$$(J, Z) = \lim_{k \to \infty} (I^{n_k}, Z^{n_k}) = 0.$$

Therefore J is a current generated by $\imath$. By Fatou's lemma we obtain

(3.26) $$\mathcal{W}(J) \leq \liminf_{k \to \infty} \mathcal{W}(I_{n_k}) \leq \limsup_{k \to \infty} \mathcal{W}(I_{n_k}) \leq \mathcal{W}(I_M).$$

By the minimality of I_M we have also $\mathcal{W}(I_M) \leq \mathcal{W}(J)$. Therefore $\mathcal{W}(I_M) = \mathcal{W}(J)$, whence $J = I_M$ by Theorem (3.21). Hence

$$\lim_{k \to \infty} \mathcal{W}(I_{n_k}) = \mathcal{W}(I_M).$$

It follows that I_{n_k} converges to I_M in norm.

The above arguments show that every subsequence of I_n must contain a further subsequence strongly convergent to I_M. Therefore $\lim_{n \to \infty} I_n = I_M$ in $\mathbf{H}_1$.

Finally, if $J \in \mathbf{H}_1$ is such that $\partial J = -\imath$, then the restriction J^n of J to Γ'_n satisfies $\partial_n J_n = -\imath_n$. Hence J_n has energy greater than $\mathcal{W}(I'_n)$, by Thomson's principle for finite networks. Letting $n \to \infty$ we get $\mathcal{W}(I_M) \leq \mathcal{W}(J)$. The functional $\mathcal{W}$ is strictly convex, so that the minimum point on the closed convex set of all flows generated by $\imath$ is unique. Hence, either $J = I_M$ or $\mathcal{W}(I_M) < \mathcal{W}(J)$. $\square$

REMARK. The limit current is obtained by passing to the limit on the currents on finite *open* subnetworks, while the minimal current arises is the limit of currents in finite subnetworks obtained by shorting the nodes outside finite subsets. In this sense, the minimal current is obtained by "shorting the network at infinity".

The last statement in Theorem (3.25) can be cosidered as the extension to infinite networks of Thomson's principle.

Corollary (3.27). *Suppose that $\imath$ is a 0-chain and that I'_n is the current generated by $\imath_n$ in (Γ'_n, r'_n). Then $\imath \in \partial\mathbf{H}_1$ if and only if $\mathcal{W}(I'_n)$ is bounded.*

Proof. The "if" part is contained in the statement of Theorem (3.25). The "only if" part can be proved by observing that, as in the proof of Theorem (3.25), the boundedness of the sequence of the energies $\mathcal{W}(I'_n)$ implies that some subsequence of I_n converges to a current $I \in \mathbf{H}_1$ generated by $\imath$. The conclusion follows from Lemma (3.3). □

DEFINITION. Let $\imath = \delta_b - \delta_a$ be a dipole. Let u_M denote the potential of the minimal current generated by $\imath$. The number

$$R_M = R_M(a, b) = u_M(a) - u_M(b)$$

is called the (minimal) effective resistance of the network between a and b.

We have the analogue of Corollary (3.20) for the minimal current.

Corollary (3.28). *Let I_M be the minimal current generated by the dipole $\delta_b - \delta_a$, and let $u_M = dI_M$ be the corresponding potential. Then*

$$u_M(a) - u_M(b) = R_M(a, b) = \mathcal{W}(I_M).$$

Proof. With the same notation as in Theorem (3.25), let u'_n be the potential of I'_n which is zero at a fixed vertex $q \in V_1$. Then, for every edge $[x, y] \in Y$ we have

$$i(x, y) = \lim_{n \to \infty} i'_n(x, y) = \lim_{n \to \infty} c(x, y)(u'_n(x) - u'_n(y)).$$

Hence $u'_n(x) - u'_n(y)$ converges to $r(x, y)i(x, y)$. Since $u'_n(q) = 0$ for all n, $u'_n(x)$ converges for every x to $u(x)$, the potential of I which is zero at q.

By the law of the conservation of the energy for finite networks we have

$$\mathcal{W}(I'_n) = R'_n = u_n(a) - u_n(b),$$

where R'_n is the effective resistance between a and b in (Γ'_n, r'_n). Letting n tend to infinity we get the thesis from Theorem (3.25). □

In order to make full use of Theorem (3.25) it is necessary to express the energy $\mathcal{W}(I'_n) = \mathcal{W}(I_n)$ of the finite currents I'_n in terms of their potentials.

Let u'_n denote the potential of I'_n which vanishes at b_n. Let P'_n denote random walk operator on (Γ'_n, r'_n). Denote by $G'_n(x, y) = G'(x, y; b_n)$ the expected number of visits to y before hitting b_n, when the random walk starts at x $(x, y \in V'_n)$. By

Theorem (2.7), $u'_n(x) = G'_n(f)(x)$, where $f(y) = -c(y)^{-1}\imath(y)$. Since $c(x) = c'_n(x)$ for $x \neq b_n$, we have

$$(3.29) \quad \mathcal{W}(I'_n) = D(G'_n(f)) = \frac{1}{2} \sum_{x,y \in V'_n} c'_n(x) p'_n(x,y) \big((G'_n(f)(x) - G'_n(f)(y) \big)^2 =$$

$$= \sum_{x,y \in V'_n} c'_n(x) p'_n(x,y) G'_n(f)(x) \big((G'_n(f)(x) - G'_n(f)(y) \big) =$$

$$= \sum_{x \in V_n} c(x) G'_n(f)(x) [(1 - P'_n)(G'_n(f))(x)] = \sum_{x \in V_n} c(x) G'_n(f)(x) f(x) =$$

$$= \sum_{x,y \in V_n} c(x) G'_n(x,y; b_n) f(x) f(y).$$

By (3.29), Theorem (3.25) and Corollary (3.27) a current generated by $\imath$ exists in $\mathbf{H}_1$ if and only if $\sum_{x,y \in V_n} c(x) G'_n(x,y; b_n) f(x) f(y)$ is bounded as $n \to \infty$.

We continue our study of the minimal current by examining separately the transient and the recurrent case.

§4. Transient networks

If (Γ, r) is a transient network and $G(|f|)(x)$ is finite for all x, then, by Lemma (1.32), $G(f)(x)$ is a solution of Poisson's equation. The following Theorem shows that if condition (3.31) below holds, then $G(f)$ is in fact the potential of the minimal current (Schlesinger (1992)).

Theorem (3.30). *Suppose that* (Γ, r) *is a transient network. Let* $\imath$ *be a 0–chain and set* $f(x) = -c(x)^{-1}\imath(x)$. *A sufficient condition in order that* $\imath \in \partial \mathbf{H}_1$ *is that*

$$(3.31) \qquad \sum_{x,y \in V} c(x) G(x,y) |f(x) f(y)| < \infty.$$

This condition is also necessary when f *is nonnegative.*

If (3.31) holds and I_M *is the minimal current generated by* $\imath$, *then*

   (1) $G(|f|)(x) = \sum_{y \in V} G(x,y)|f(y)| < \infty$ *for all* $x \in V$;

   (2) $G'_n(f)(x) \to G(f)(x)$ *for all* x;

   (3) $G(f)$ *is the potential of* I_M;

   (4) $\mathcal{W}(I_M) = \sum_{x,y \in V} c(x) G(x,y) f(x) f(y)$.

Proof. By Corollary (3.27) and equations (3.29) there exists a current generated by $\imath$ if and only if

$$D(G'_n(f)) = \sum_{x,y \in V_n} c(x) G'_n(x,y; b_n) f(x) f(y)$$

is bounded.

Let $\{p'(x,y;b_n)\}$ denote the Markov chain associated with (Γ'_n, r'_n) where b_n has been made absorbing. Let $p'^k(x,y;b_n)$ denote the corresponding k-th step transition probabilities. For every n and $k \geq 0$, and for every $x, y \in V_n$

$$p'^{k+1}(x,y;b_n) \leq p'^k(x,y;b_{n+1}) \leq p^k(x,y).$$

Moreover $p'^k_n(x,y)(b_n) \to p^k(x,y)$ as $n \to \infty$ for all k. Therefore $G'_n(x,y;b_n)$ increases to $G(x,y)$ for all $x, y \in V$. Hence we have

$$D(G'_n(f)) = \sum_{x,y \in V_n} c(x)G'_n(x,y;b_n)f(x)f(y) \leq \sum_{x,y \in V} c(x)G(x,y)|f(x)f(y)| < \infty.$$

for all n. It follows that $\imath \in \partial \mathbf{H}_1$.

If f is nonnegative, then $D(G'_n(f))$ tends to $\sum_{x,y \in V} c(x)G(x,y)f(x)f(y)$ by the monotone convergence theorem, so that (3.31) is also necessary.

To prove assertion (1) we may assume that $\imath(x_0) \neq 0$ for some $x_0 \in V$. Then, (3.31) implies that $G(|f|)(x_0) < \infty$. For every $x \in V$ (and large values of n) we have

$$|G'_n(|f|)(x) - G'_n(|f|)(x_0)| \leq KD(G'_n(|f|))^{1/2}$$
$$\leq K\Big(\sum_{z,y} c(z)G(z,y)|f(z)f(y)|\Big)^{1/2}.$$

where K is a constant depending only on x and x_0 (compare with Lemma (3.14)). By the monotone convergence theorem $G'_n(|f|)(x) \uparrow G(|f|)(x)$. Since $G(|f|)(x_0) < \infty$, we obtain $G(|f|)(x) < \infty$ for all x. Hence (1) holds.

By (1) and the dominated convergence theorem we have for all $x \in V$

$$\sum_{y \in V'_n} G'_n(x,y;b_n)f(y) \to \sum_{y \in V} G(x,y)f(y) \qquad \text{as } n \to \infty.$$

Hence (2) holds. By Fatou's Lemma and Theorem (3.25)

$$D(G(f)) \leq \lim_{n \to \infty} D(u_n) = \lim_{n \to \infty} \mathcal{W}(I_n) = \mathcal{W}(I_M).$$

Since $G(f)$ is a potential it follows that $G(f)$ must be the potential of the minimal current. In other words $D(G(f)) = \mathcal{W}(I_M)$ so that (3) and (4) hold. $\square$

Corollary (3.32). *If (Γ, r) is a transient network, then for every $a \in V$ the 1-chain $\imath = -c(a)\delta_a$ is an element of $\partial \mathbf{H}_1$. The potential of the minimal current generated by $\imath$ is $G(\cdot, a)$. Moreover $G(\cdot, a)$ belongs to $\mathbf{D}_0$ for every $a \in V$.*

Proof. We have $f = \delta_a$. Then $G(f)(x) = G(x, a)$ for all x. By Theorem (3.30) $\imath = -c(a)\delta_a$ is in $\partial \mathbf{H}_1$, and the potential of the minimal current generated by $\imath$ is $u_M(x) = G(x, a)$.

Let (Γ'_n, r'_n) be as in Theorem (3.25), and suppose $a \in V'_1$. Set, for every $x \in V$, $u_n(x) = G'(x, a; b_n)$ if $x \in V_n$, $u_n(x) = 0$ if $x \notin V_n$. Then, $u_n \in \mathbf{D}_0$ and $D(u_n) = \mathcal{W}(I_n) \leq \mathcal{W}(I_M)$. Therefore, passing to a subsequence if necessary, u_n is weakly convergent to some $u \in \mathbf{D}_0$. By Lemma (3.15) $u_n(x) \to u(x)$ for all x. On the other hand, by (2) and (3) of Theorem (3.30), u_n tends to u_M pointwise. Hence $u_M = u$ belongs to $\mathbf{D}_0$. $\square$

Thus in the transient case, unlike the finite case, there are always currents generated by finite energy sources which are not "balanced", in the sense that $\sum_x \imath(x) \neq 0$. For instance $\imath = -c(a)\delta_a$. This fact can be interpreted in the following way: the current enters the network at a and leaves the network "at infinity". We will see in the next section that this is not the case when the network is recurrent.

We can deduce from the above results the well known "simple criterion" of transience of T. Lyons (1983) (see Kalpazidou (1991) for another proof).

Theorem (3.33). *A necessary and sufficient condition for the transience of a network is that there exists a vertex $a \in V$ and a 1-chain $I \in \mathbf{H}_1$ such that $\partial I + \delta_a = 0$.*

Proof. If the network is transient, then we apply Corollary (3.32). Conversely, assume that $-c(a)\delta_a \in \partial \mathbf{H}_1$ for some $a \in V$. Let (Γ'_n, r'_n) be as Theorem (3.25). Set, as above, $u_n(x) = G'(x, a; b_n)$ if $x \in V_n$, $u_n(x) = 0$ if $x \notin V_n$. Then, u_n is bounded in $\mathbf{D}$ by Corollary (3.27). Therefore, the increasing sequence $G'(x, a; b_n)$ is pointwise convergent by Theorem (3.15). As $G'(x, a; b_n)$ increases to $G(x, a)$ for all x, we may conclude that the network is transient. $\square$

Let (Γ, r) be a network. Choose any $a \in V$ and let (Γ_n, r_n) be any exhaustion of (Γ, r) such that $a \in V_1$. Denote by $R_n = c(a)^{-1} G_n(a, a; b_n)$ the effective resistance of the network (Γ'_n, r'_n) between a and b_n. We already observed that the sequence R_n increases monotonically and that its limit is $c(a)^{-1} G(a, a)$.

DEFINITION. Let (Γ, r) be an infinite network and let $a \in V$. The limit

$$R(a) = \lim_{n \to \infty} R_n = c(a)^{-1} G(a, a)$$

is called the effective resistance of the network between a and ∞. Its inverse, $C(a) = R^{-1}(a)$ is called the effective conductance between a and ∞.

REMARK. A network is transient if and only if $R(a) < \infty$ for some (=all) $a \in V$. In this case, by Theorem (3.30), $R(a)$ is the potential of the minimal current generated by the 0–chain $\imath = -\delta_a$.

If (Γ, r) is a transient network and $J \in \mathbf{H}_1$ is such that $\partial J + \delta_a = 0$, then $R(a) \le \mathcal{W}(J)$. This is an immediate consequence of Corollary (3.32) and Theorem (3.25).

§5. Recurrent networks

We already proved (Theorem (1.33)) that a recurrent network does not have nonconstant positive superharmonic functions. We can now prove that a recurrent network has no nonconstant superharmonic functions with finite Dirichlet sums.

Theorem (3.34). *Let (Γ, r) be a recurrent network. Suppose that $u \in \mathbf{D}$ satisfies $u(x) \ge P(u)(x)$ (or $u(x) \le P(u)(x)$) for all $x \in V$. Then u is constant. In particular, any 0–chain $\imath \in \partial \mathbf{H}_1$ generates a unique current in the network.*

Proof. First we prove that every superharmonic function with finite Dirichlet sum must be harmonic.

Suppose e.g. that $u \in \mathbf{D}$ is subharmonic and let $f(x) = u(x) - P(u)(x)$. Then f is a non–positive function and u is the potential of a current generated by $\imath$, where $\imath(x) = -c(x)f(x) \ge 0$. If f is not identically 0, let I'_n be as in Theorem (3.25). We have

$$\mathcal{W}(I'_n) = \sum_{x,y \in V_n} G'_n(x, y; b_n) c(x) f(x) f(y) \to \infty$$

since $G'_n(x, y; b_n)$ tends monotonically to $G(x, y) = \infty$. By Corollary (3.27) $\imath \notin \partial \mathbf{H}_1$, so that $u \notin \mathbf{D}$. Thus $u(x) = P(u)(x)$ for all x.

Let now u be a harmonic function with finite Dirichlet sum. Let $u^+ = \max(u, 0)$ and $u^- = \max(-u, 0)$. Since $|u^+(x) - u^+(y)| \le |u(x) - u(y)|$ and $|u^-(x) - u^-(y)| \le |u(x) - u(y)|$, we have that u^+ and u^- belong to $\mathbf{D}$. Moreover

$$u(x) \le \sum_y p(x, y) \max(u(y), 0),$$

so that $u^+(x) \le P(u^+)(x)$. In the same way we see that $u^-(x) \le P(u^-)(x)$. By what we have already proved u^+ and u^- must be harmonic. By Theorem (1.33) they are constant. $\square$

Corollary (3.35). *Let* (Γ, r) *be a recurrent network. Suppose that* $\imath \in \partial \mathbf{H}_1$ *and set* $f(x) = -c(x)^{-1}\imath(x)$. *Let* q *be any fixed vertex in* V. *If* $u \in \mathbf{D}$ *satisfies* $(1 - P)(u)(x) = f(x)$ *for all* $x \neq q$, *then* u *satisfies also* $(1 - P)(u)(q) = f(q)$.

Proof. Let $\bar{u}$ be the potential of the current generated by $\imath$. Then $v = u - \bar{u}$ is subharmonic or superharmonic on V. Hence $v = $ const. $\square$

If a network is transient, then by Theorem (3.30) the potential of the minimal current generated by $\imath \in \partial \mathbf{H}_1$ can be expressed as $G(f)$. In the recurrent case this is not possible, as $G(f)$ is undefined.

The minimal current is expressed in any case by Theorem (3.25) as a limit of currents in finite networks obtained by shorting the vertices not in V_n. It will be shown below that the (unique!) current with finite energy in a recurrent network is also the limit of currents in finite network obtained by shorting together the vertices outside V_n with a fixed node $q \in V_1$.

Suppose that (Γ_n, r_n) is an exhaustion of (Γ, r). Let $\Gamma_n = (V_n, Y_n)$ and fix $q \in V_1$. For every n we define a new network $(\Gamma_n'^q, r_n')$ by shorting together q and all the vertices in V which are not in V_n. Denote by b_n the vertex of $\Gamma_n'^q$ that is not in $V_n \setminus \{q\}$ i.e., the vertex obtained by contracting q and all the vertices in $V \setminus V_n$. Let V_n' and Y_n' denote the vertex and edge set of $\Gamma_n'^q$ respectively. Clearly $V_n' = (V_n \setminus \{q\}) \bigcup \{b_n\}$.

The graph in Figure 3.2 is obtained by shorting together the endpoints of the dashed edges in Figure 3.1. The graph in Figure 3.3 is obtained by shorting the same vertices with q. The dotted edges in Figure 3.3 are identified with the solid edges having the same endpoints.

The conductances $c(x, y)$ do not change if $x, y \in V_n \setminus \{q\}$. Hence, for all $x \in V_n \setminus \{q\}$, $c(x)$ is the same in (Γ, r) and in $(\Gamma_n'^q, r_n')$.

For every n we define a 0–chain $\imath_n$ on $\Gamma_n'^q$ in the following way

$$\imath_n(x) = \begin{cases} \imath(x) & \text{if } x \in V_n \setminus \{q\} \\ -\sum_{y \in V_n \setminus \{q\}} \imath(x) & \text{if } x = b_n. \end{cases}$$

Note that $\imath_n$ does not depend on the value of $\imath$ in q. The following Theorem, contained in Schlesinger (1992), is the version of Theorem (3.25) for recurrent networks.

Theorem (3.36). *Suppose that* (Γ, r) *is a recurrent network and let* $\imath \in \partial \mathbf{H}_1$. *Let* I_n' *be the current generated by* $\imath_n$ *in* $(\Gamma_n'^q, r_n')$ *and let* I_n *be its extension to* Γ. *Then* $\mathcal{W}(I_n')$ *is a bounded sequence and* I_n *converges in* $\mathbf{H}_1$ *to the unique current* I *generated by* $\imath$.

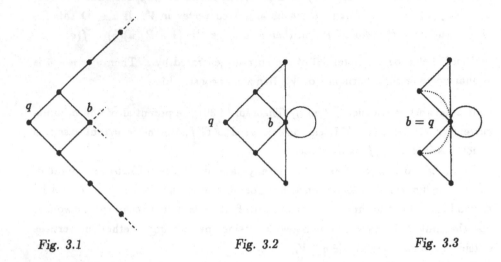

Fig. 3.1   Fig. 3.2   Fig. 3.3

Proof. The proof is similar to the proof of Theorem (3.25). Let I be the current generated by $\imath$ in (Γ, r) and let I^n be the restriction of I to $\Gamma_n'{}^q$. Then, $\mathcal{W}(I^n) \leq \mathcal{W}(I)$. If ∂_n is the boundary operator on $\Gamma_n'{}^q$, then $(\partial_n I^n)(x) = (\partial I)(x)$ at every $x \in V_n \setminus \{q\}$. Since $\partial I + \imath = 0$, by Lemma (2.1) and the definition of $\imath_n$, $\partial_n I^n(x) + \imath_n(x) = 0$ for every $x \in V_n'$. Therefore I^n is a flow generated by $\imath_n$.

Let I_n' denote the current generated by $\imath_n$ in $(\Gamma_n'{}^q, r_n')$, and let I_n be its extension to Γ. By Thomson's principle for finite networks

$$\mathcal{W}(I_n) = \mathcal{W}(I_n') \leq \mathcal{W}(I^n) \leq \mathcal{W}(I).$$

Therefore $\mathcal{W}(I_n) = \mathcal{W}(I_n')$ is bounded.

There exists a subsequence I_{n_k} weakly convergent in $\mathbf{H}_1$ to a 1–chain J. We claim that J is the current I generated by $\imath$.

The value of I_{n_k} on every branch $B \in Y$ converges to the value of J on B. By Lemma (3.3) $\partial I_{n_k}(x)$ converges to $\partial J(x)$ for every $x \in V \setminus \{q\}$. If k is sufficiently large $(\partial I_{n_k})(x) = -\imath(x)$. Therefore $(\partial J)(x) + \imath(x) = 0$ for every $x \neq q$.

Let now Z be a finite cycle on Γ. First suppose that the value of Z on every edge of the form $[x, q]$ is zero. Then, for large values of n, the restriction Z^n of Z to $\Gamma_n'{}^q$ is a cycle. If n is large enough r and r_n' have the same values on the edges B such that $z(B) \neq 0$. Therefore $0 = (I_{n_k}', Z^{n_k}) = (I_{n_k}, Z)$. Letting k tend to infinity we obtain $(J, Z) = 0$.

Suppose now that, for some $x \sim q$, $z(x, q) \neq 0$. Let $Z^n = \{z^n(B)\}$ be as before the restriction to $\Gamma_n'{}^q$. If n is large enough, then all the edges $B \in Y$ such that q

is not an endpoint of B and $z(B) \neq 0$ are in Y_n. For such edges $z^n(B) = z(B)$, and $r(B) = r'_n(B)$. If $x \sim q$ and $z(x,q) \neq 0$, then, always for large values of n, $B = [x,q]$ is the unique edge such that $z(B) \neq 0$ and $\Phi_n(B) = [x,b_n]$. Therefore $z^n(x,b_n) = z(x,q)$. It follows, for n sufficiently large,

$$\partial_n Z^n(b_n) = \partial Z(q) = 0,$$

so that Z^n is a cycle on $\Gamma'_n{}^q$.

Denote by $i'_n(B')$ and $i_n(B)$ the values of I'_n and I_n on B' and B respectively. By the definition of extension we have $i_n(x,q)r(x,q) = i'_n(x,b_n)r'_n(x,b_n)$ for all $x \sim q$. Hence, for large values of n,

$$0 = (I'_n, Z^n) = \sum_{B' \in Y'_n} i''_n(B')z^n(B')r'_n(B')$$

$$= \sum_{B \in Y} z(B)i_n(B)r(B) = (I_n, Z).$$

Letting n tend to infinity, we see that J satisfies Kirchhoff's loop law. Therefore there exists a function $u \in \mathbf{D}$ such that $du = J$. Since J satisfies also Kirchhoff's node law at every point $x \neq q$, we have $(1 - P)(u)(x) = f(x)$ (where $f(x) = c(x)^{-1}\iota(x)$) for every $x \neq q$. By Corollary (3.35) u satisfies Poisson's equation everywhere. Hence J is the current generated by ι. The proof that I_n converges strongly to I now is as in Theorem (3.25). $\square$

We have the analogue of Corollary (3.27).

Corollary (3.37). *Suppose that ι is a 0–chain and that I'_n is the current generated by ι_n in $(\Gamma'_n{}^q, r'_n)$. Then $\iota \in \partial \mathbf{H}_1$ if and only if $\mathcal{W}(I_n)$ is bounded.*

For all $q \in V$ and every function f set $G(f;q)(x) = \sum_y G(x,y;q)f(y)$ whenever this expression is defined. We have the recurrent analogue of Theorem (3.30) in which the function $G(x,y)$ is replaced by $G(x,y;q)$.

Theorem (3.38). *Let (Γ, r) be a recurrent network and let q be any vertex in V. For every 0–chain ι set $f(x) = -c(x)^{-1}\iota(x)$. Suppose that*

$$(3.39) \qquad \sum_{x,y \in V} c(x)G(x,y;q)|f(x)f(y)| < \infty$$

Then $\iota \in \partial \mathbf{H}_1$. Furthermore

(1) $G(|f|;q)(x) = \sum_{y \in V} G(x,y;q)|f(y)| < \infty$ *for all $x \in V$;*

(2) $G'_n(f;q)(x) \to G(f;q)(x)$ *for all x;*

(3) $G(f;q)$ *is the potential of I;*

(4) $\mathcal{W}(I) = \sum_{x,y \in V} c(x)G(x,y;q)f(x)f(y),$

where I the current generated by $\imath$.

Proof. Analogous to the proof of Theorem (3.30). □

Theorem (3.40). *Suppose that* $\imath = \sum_x \imath(x)\delta_x$ *is a 0–chain on a recurrent network and let* $f(x) = c(x)^{-1}\imath(x)$. *Let* $q \in V$. *Then,* $\sum_x |\imath(x)| < \infty$ *implies* $G(|f|;q)(x) < \infty$ *for all* x. *If* $\deg(q) < \infty$, *then the converse is also true.*

Proof. Suppose $\sum_y |\imath(y)| = \sum_y c(y)|f(y)| < \infty$. Then, (we may assume $x \neq q$),

$$\sum_{y \neq q} G(x,y;q)|f(y)| = \sum_y G(y,x;q)c(x)^{-1}c(y)|f(y)| =$$

$$c(x)^{-1}G(x,x;q)\sum_y c(y)H(y,x;q)|f(y)| \leq c(x)^{-1}G(x,x;q)\sum_y c(y)|f(y)| < \infty.$$

To prove the reverse implication we use again the fact that $x,y \neq q$ implies $c(x)G(x,y;q) = c(y)G(y,x;q)$. Therefore we obtain for all $y \neq q$

$$\sum_z c(q)p(q,x)G(x,y;q) = c(y)\sum_{x\neq q} p(x,q)G(y,x;q)$$

$$= c(y)\sum_{n=0}^{\infty}\sum_{x\neq q} p^n(y,x;q)p(x,q).$$

As

$$\sum_{n=0}^{\infty}\sum_{x\neq q} p^n(y,x;q)p(x,q) = \overline{H}(y,q) = 1,$$

we obtain

$$\sum_z c(q)p(q,x)G(x,y;q) = c(y) \quad \text{for all } y \neq q.$$

It follows

$$\sum_{y\neq q} |\imath(y)| = \sum_{y\neq 0}\sum_z c(q)p(q,x)G(x,y;q)c(y)^{-1}|\imath(y)|$$

$$= \sum_{x\sim q} c(q,x)G(|f|;q)(x) >$$

Thus $\sum_y |\imath(y)| < \infty$ if $\deg(q) < \infty$. □

§6. Dirichlet's and Rayleigh's principles

In §3 we have shown that the minimal current generated by a 0–chain with finite energy is the flow with minimum energy (Theorem (3.25)). A version of Dirichlet's and Rayleigh's principles can also be proved in the infinite case. We start with Dirichlet's principle.

Theorem (3.41). *Let (Γ, r) be an infinite network and let a be any vertex in V. Let $R(a)$ denote the effective resistance between a and ∞. Then*

$$R^{-1}(a) = \inf\{D(u) : u \in \ell_0, u(a) = 1\}.$$

Proof. Let (Γ_n', r_n') be as in Theorem (3.25). Fix $\epsilon > 0$ and choose $v \in \ell_0$ in such a way that

$$D(v) < \inf\{D(u) : u \in \ell_0, u(a) = 1\} + \epsilon.$$

There exists n_0 such that $v(x) = 0$ for $x \notin V_{n_0}$. Therefore, for all $n \geq n_0$, v can be identified with a function on V_n' such that $v(b_n) = 0$.

By Theorem (2.16) $c(a)G(a, a; b_n)^{-1} \leq D(v)$. Hence

$$R^{-1}(a) = c(a)G(a, a)^{-1} = c(a) \lim_{n \to \infty} G(a, a; b_n)^{-1} \leq \inf\{D(u) : u \in \ell_0, u(a) = 1\}.$$

Now we show that $R^{-1}(a)$ is actually the infimum. Suppose that v_n is the potential, with $v_n(b_n) = 0$, of the current generated by $-c(a)G(a, a; b_n)^{-1}(\delta_{b_n} - \delta_a)$ on (Γ_n', r_n'). Define the function u_n on V by setting $u_n(x) = v_n(x)$ if $x \in V_n$, $u_n(x) = 0$ otherwise. Then, $u_n \in \ell_0$ and, by (2.9), $u_n(a) = 1$. By Theorem (2.16) $D(u_n) = D(v_n) = c(a)G(a, a; b_n)^{-1} \to R^{-1}(a)$ as $n \to \infty$. $\square$

Since $R = \infty$ if and only if the network is recurrent, we get the following Corollary.

Corollary (3.42). *An infinite network (Γ, r) is recurrent if and only if for some $(=$all$)$ $a \in V$ there exists a sequence of finitely supported functions u_n on V such that $u_n(a) = 1$ and $\lim_{n \to \infty} D(u_n) = 0$.*

Another consequence of Dirichlet's principle is Lyons' "better test" of transience (Yamasaki (1982), T. Lyons (1983)). Contrary to the transient case, a summable 0–chain on a recurrent network generates a current only if it is "balanced".

Theorem (3.43). *Let (Γ, r) be a recurrent network and let $\imath$ be a 0–chain on Γ. If $\sum_x |\imath(x)| < \infty$ and $\imath \in \partial H_1$, then $\sum_x \imath(x) = 0$.*

Proof. Choose any vertex $a \in V$. By Corollary (3.42) there is a sequence of finitely supported functions u_n such that $u_n(a) = 1$ and $D(u_n) \to 0$ as $n \to \infty$.

Setting

$$g_n(x) = \min\big(1, \max(0, u_n(x))\big), \quad \cdot$$

we have that $g_n(a) = 1$, $0 \le g_n \le 1$ and $D(g_n) \le D(u_n) \to 0$. In particular, $|g_n(x) - g_n(y)| \to 0$ as $n \to \infty$ for all $x \sim y$. As $g_n(a) = 1$, $g_n(x) \to 1$ for every $x \in V$ by connectedness.

Let $I \in \mathbf{H}_1$ be such that $\partial I = \imath$. Then, since g_n is finitely supported,

$$\sum_{x \in V} \imath(x)g_n(x) = \sum_{x \in V} \partial I(x)g_n(x)$$

$$= \frac{1}{2} \sum_{[x,y] \in Y} i(x,y)\big(g_n(x) - g_n(y)\big)$$

$$= (I, dg_n) \to 0.$$

By the dominated convergence theorem

$$0 = \lim_{n \to \infty} \sum_{x \in V} \imath(x)g_n(x) = \sum_{x \in V} \imath(x).$$

$\square$

We come now to Rayleigh's principle for infinite networks. We will present a general version of such a principle in terms of morphisms between networks.

Let (Γ, r) and (Γ', r') be networks and let ϕ be a network morphism from (Γ, r) to (Γ', r') with the property that

(3.44) $\mathrm{card}\{s \in V : \phi(s) = x\} < \infty \quad$ for all $x \in \phi(V)$.

Let $\imath$ be a 0–chain on Γ. Then we may define a 0–chain $\imath'$ on Γ' by setting for all $x \in V'$

$$\imath'(x) = \begin{cases} \displaystyle\sum_{\phi(s)=x} i(s) & \text{if } x \in \phi(V) \\[2mm] 0 & \text{otherwise.} \end{cases}$$

Theorem (3.45). *Let (Γ, r) and (Γ', r') be networks and suppose that ϕ is a network morphism from (Γ, r) to (Γ', r') satisfying (3.44). Let $i \in \partial\mathbf{H}_1$ be a 0–chain and define $\imath'$ on Γ' as above. Then $\imath'$ is the boundary of a 1–chain with finite energy on (Γ', r). Denoting by I and I' the minimal currents generated by i and $\imath'$ respectively, we have*

$$\mathcal{W}(I') \le \mathcal{W}(I).$$

Proof. We define a 1–chain J on Γ' in the following way:

$$j(B') = \sum_{\{B \in \Phi^{-1}(B')\}} i(B) \quad \text{if } B' \in \Phi(Y) \text{ is not a self–loop}$$

$$= 0 \qquad\qquad \text{otherwise.}$$

The 1–chain J has finite energy since

$$|j(B')|^2 \leq \sum_{\{B \in \Phi^{-1}(B')\}} r(B)^{-1} \sum_{\{B \in \Phi^{-1}(B')\}} r(B)|i(B)|^2$$

$$\leq \frac{1}{r(B')} \sum_{\{B \in \Phi^{-1}(B')\}} r(B)|i(B)|^2 .$$

Summing over B' we obtain $\mathcal{W}(J) \leq \mathcal{W}(I)$. It is easily seen that $\partial' J' = -\imath'$. Namely, for every $x \in \phi(V)$,

$$\partial' J(x) = \sum_{y \in V'(x)} j(x,y) = \sum_{\phi(s)=x} \sum_{t \sim s} i(s,t)$$

$$= \sum_{\phi(s)=x} \partial I(s) = - \sum_{\phi(s)=x} i(s)$$

$$= -\imath'(x).$$

On the other hand, if $x \notin \phi(V)$, then $j(x,y) = 0$ for all $y \in V'$, $y \sim x$, so that $\partial' J(x) = 0$. Hence $\imath' \in \partial H_1$. Let I' be the minimal current generated by $\imath'$. By the minimality of I' we have $\mathcal{W}(I') \leq \mathcal{W}(J) \leq \mathcal{W}(I)$. □

DEFINITION. Let (Γ, r) be a network and let U be a (infinite) subset of Γ. Suppose that

$$(3.46) \qquad\qquad \sum_{x \in U} \sum_{\substack{y \in V(x) \\ y \notin U}} \frac{1}{r(x,y)} < \infty.$$

Then r' satisfies (1.1) and the network (Γ', r'), obtained by shorting together all the nodes of U, is well defined. In this case we have another version of Rayleigh's principle which is not contained in Theorem (3.45), even though the arguments of the proof are essentially the same.

Theorem (3.47). *Let (Γ, r) be a network and let $U \subseteq V$ be a subset of V such that (3.46) holds. Suppose that (Γ', r') is obtained by shorting together the nodes in U. Let $\imath$ be a finite 0–chain such that $\partial \imath = 0$. Let $\imath'$ be the 0–chain on Γ' defined* as in Theorem (3.45). Then, denoting by I and I' the currents generated by $\imath$ in (Γ, r) and (Γ', r') respectively, we have $\mathcal{W}(I') \leq \mathcal{W}(I)$.

Proof. We repeat the proof of Theorem (3.45) taking (3.46) into account. Note that J is well defined, since for every $x \notin U$,

$$
\begin{aligned}
|j(x,b)| &\leq \sum_{\substack{z \in U \\ z \sim z}} |i(x,z)| \\
&\leq \sum_{\substack{z \in U \\ z \sim z}} c(x,z) \sum_{\substack{z \in U \\ yzsimz}} r(x,z) i^2(x,z).
\end{aligned}
$$

Analogously, $\partial' J'(x) = -\imath'$. $\square$

Since the effective resistance between two points or between a point and infinity is the energy of the current generated by a dipole or by a 0–chain with mass only at one point, we can deduce a more familiar form of Rayleigh's principle.

Corollary (3.48).

(1) If $r_1(B) \leq r_2(B)$ for all B then, the effective resistance between two points or between a point and infinity in (Γ, r_1) is not greater than the same resistance in (Γ, r_2). In particular (Γ, r_2), is recurrent if so is (Γ, r_1).

(2) Suppose that (Γ_2, r_2) is subnetwork of (Γ_1, r_1). Let a, b be vertices of Γ_2. Then the effective resistance between a and b in the subnetwork is not smaller than the same resistance in the network. The same is true for the resistance between a and infinity. In particular, a subnetwork of a recurrent network is recurrent.

Corollary (3.49). Let $V_n, j = 1, 2, \ldots$ be a countable family of pairwise disjoint finite subsets of V. Assume that a and b do not belong to the same set V_n. Then, shorting together the nodes in V_n for all n decreases (or does not affect) the effective resistance between a and b and between a and ∞. Cutting one or more edges (possibly infinitely many!) increases (or does not affect) the effective resistance between a and b and the effective resistance between a and ∞.

DEFINITION. Let $\Gamma = (V, Y)$ be an infinite graph. A sequence of finite subsets $V_n \subseteq V$ is called a canonical decomposition of Γ if they have the following properties:

(1) $V = \bigcup_{n=0}^{\infty} V_n$;

(2) $V_n \cap V_m = \emptyset$ if $n \neq m$;

(3) $x \in V_n$ and $x \sim y$ together imply $y \in V_{n-1} \cup V_n \cup V_{n+1}$ $(V_{-1} = \emptyset)$.

The following result is the well known Nash–Williams' sufficient condition for recurrence (Nash–Williams (1959)).

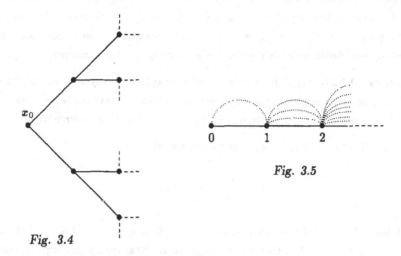

Fig. 3.5

Fig. 3.4

Theorem (3.50). *Let* (Γ, r) *be a locally finite network. Suppose that there exists a canonical decomposition* $\{V_n\}$ *of* Γ *such that*

(3.51)
$$\sum_{n=1}^{\infty} \frac{1}{\sum_{x \in V_{n-1}, y \in V_n} c(x,y)} = \infty.$$

Then the network is recurrent.

Proof. Let Γ' be the one way path whose vertices are indexed by the non–negative integers. Set

$$r'(n-1,n) = \frac{1}{\sum_{x \in V_{n-1}, y \in V_n} c(x,y)} \quad \text{for } n = 1, 2 \ldots.$$

Up to self–loops, the network (Γ', r') is obtained by shorting the vertices in V_n for all n. The paths $\mathbf{p}_n$ joining 0 to n are an exhaustion of Γ' and coincide with the graphs Γ'_n of Theorem (3.25). The resistance R' between 0 and ∞ is the limit of the resistances R'_n between 0 and $b_n = n$ of $\mathbf{p}_n$. Clearly,

$$R'_k = \sum_{n=1}^{k} \frac{1}{\sum_{x \in V_{n-1}, y \in V_n} c(x,y)}.$$

Hence by (3.51) (Γ', r') is recurrent. Theorem (3.45) (or Corollary (3.49)) implies that (Γ, r) is recurrent too. $\square$

REMARK. We notice that the original Theorem of Nash–Williams contains also a necessary condition for recurrence, which is of less immediate application. An extension of Nash–Williams theorem to locally infinite networks is due to McGuinness (1991).

Figure 3.4 represents a binary tree Γ with root x_0 (i.e., $\deg(x_0) = 2$ and $\deg(x) = 3$ for all the other vertices x). Setting $V_n = \{x : d(x, x_0) = n\}$, the graph Γ' obtained by shorting all the vertices in V_n, $n = 1, 2 \ldots$, is represented in Figure 3.5. The dotted edges are identified with the solid edge having the same endpoints.

Corollary (3.52). *Let Γ be a locally finite graph. Fix any $o \in V$ and let e_n, $n = 1, 2, \ldots$, denote the number of edges whose endpoints have distances less than or equal to n from o. If $e_n \le \text{const.}\, n^2 \log n$ $(n \ge 2)$, then Γ is recurrent.*

Proof. By Theorem (3.50) it suffices to show that

$$(3.53) \qquad \sum_{n=1}^{\infty} \frac{1}{e_{n+1} - e_n} = \infty.$$

Without loss of generality we may assume $e_1 = 0$ and $e_n \le n^2 \log n$. Let $y(x)$ $(1 \le x < \infty)$ denote the function whose diagram is obtained by joining the points (n, e_n) and $(n+1, e_{n+1})$ by line segments. Let $x(y)$ denote the inverse function of $y(x)$ and $g(y)$ the inverse function of $f(x) = x^2 \log x$ $(x > 1)$. Then $x(y) \ge g(y)$ and $x(0) = g(0) = 1$. We have

$$(3.54) \qquad \sum_{n=1}^{\infty} \frac{1}{e_{n+1} - e_n} = \int_0^{\infty} x'^2(y)dy.$$

Set $t(y) = x(y) - g(y)$ and, for every $M > 1$,

$$F_M(\theta) = \int_0^M \left[\frac{d}{dy}(g(y) + \theta t(y)) \right]^2 dy.$$

where $0 \le \theta \le 1$. We have

$$(3.55) \qquad F_M'(\theta) = 2 \int_0^M g'(y)t'(y)dy + 2\theta \int_0^M (t'(y))^2 dy \ge 0,$$

since the first integral in (3.55) is

$$\int_0^M g'(y)t'(y)dy = t(M)g'(M) - \int_0^M t(y)g''(y)dy$$

$$= t(M)g'(M) + \int_0^M t(y) \frac{f''(x(y))}{f'(x(y))^3} dy \ge 0.$$

Thus $F_M(\theta)$ increases with θ. It follows

$$\int_0^M x'^2(y)dy \ge \int_0^M g'^2(y)dy = \int_1^{g(M)} \frac{1}{f'(x)} dx.$$

Letting M tend to infinity we get $\int_0^\infty x'^2(y)dy = \infty$. Thus we obtain (3.53) from (3.54). $\square$

EXAMPLE (3.56). The 2–dimensional grid $\mathbb{Z}^2$ is recurrent, since $e_n \leq \text{const}.n^2$. The same is true for the two way infinite path $\mathbb{Z}$.

EXAMPLE (3.57). Suppose that T is an infinite tree whose edges are assigned resistances $r(x,y)$. We suppose for simplicity that T is locally finite. Assume also that every vertex has degree greater than 1. Let $a = x_0$ be a fixed vertex. Using equation (2.12) it is easy to provide an explicit formula for the resistance R between a and ∞.

Namely, let T_n be the subtree of T induced by the vertices whose distance from a is less or equal to n. Then the family $\{T_n\}$ is an exhaustion of T, and the resistance of the network obtained shorting the vertices at distance n from a is given by (2.12). Letting n tend to infinity in (2.12) we obtain that R is expressed by the following "branching" continued fraction

(3.58)

$$R = \cfrac{1}{\displaystyle\sum_{x_1\in U(x_0)} r(x_0,x_1) + \cfrac{1}{\displaystyle\sum_{x_2\in U(x_1)} r(x_1,x_2) + \cfrac{1}{\ddots + \cfrac{1}{\displaystyle\sum_{x_n\in U(x_{n-1})} r(x_{n-1},x_n) + \ddots}}}}.$$

In particular, if the tree is homogeneous of degree q and all the resistances are 1, then we compute easily $R = (q-1)^{-1}$.

It is not easy, in general, to decide from (3.58) whether a tree is transient or recurrent, since no simple sufficient condition for the convergence of branching continued functions of type (3.58) is known; see e.g. Skorobogat'ko (1972). However, we can prove useful estimates from below and from above for the resistance R between a and infinity.

Corollary (3.59). *Let T be an infinite locally finite tree as in Example (3.57). Let V_n denote the set of all vertices having distance n from a. For every vertex x set $d_x = \deg(x)$. Then,*

(3.60)
$$\sum_{n=1}^\infty \frac{1}{\sum_{x\in V_{n-1}, y\in V_n} c(x,y)} \leq R$$

and

$$(3.61) \quad R \leq \sum_{x_1 \in U(x_0)} \frac{r(x_0, x_1)}{d_{x_0}^2} + \sum_{x_1 \in U(x_0)} \frac{1}{d_{x_0}^2} \sum_{x_2 \in U(x_1)} \frac{r(x_1, x_2)}{(d_{x_1} - 1)^2} +$$

$$+ \sum_{x_1 \in U(x_0)} \frac{1}{d_{x_0}^2} \sum_{x_2 \in U(x_1)} \frac{1}{(d_{x_1} - 1)^2} \sum_{x_3 \in U(x_2)} \frac{r(x_2, x_3)}{(d_{x_2} - 1)^2} + \cdots$$

Proof. Inequality (3.60) is a consequence of Nash–Williams Theorem (3.50), and it is obtained by shorting the vertices in V_n.

As for (3.61), we proceed as follows. Let y_j denote the neighbours of a $(j = 1, \ldots, d_a)$. Let Γ_j denote the induced subtree of T whose vertices are y_j and all the vertices x such that a does not lie in the the path joining y_j and x. Let R_j denote the resistance between y_j and ∞ in Γ_j.

Then, a convexity argument shows that

$$(3.62) \qquad R = \frac{1}{\displaystyle\sum_j \frac{1}{r(x_0, y_j) + R_j}} \leq \sum_j \frac{r(x_0, y_j) + R_j}{d_{x_0}^2}.$$

Inequality (3.61) follows by repeated applications of (3.62).   □

§7. Decomposition and approximation of functions in D

This section and the next two sections are devoted to the theory of Dirichlet finite functions on a network, which was developed in analogy with the theory of Dirichlet functions on a Riemann manifold by Yamasaki and his collaborators. The results presented in this section and in sections 8 and 9 are contained in the papers by Yamasaki (1975), (1977), (1979), (1986) and by Nakamura and Yamasaki (1977), Kayano and Yamasaki (1984). Most results proved originally for locally finite networks by Yamasaki et alii are generalized here to general countable networks; see also Soardi and Yamasaki (1993).

Theorem (3.63). *Let (Γ, r) be an infinite network and denote by $e(x)$ the function on V identically equal to 1. Then, the following are equivalent*

(1) (Γ, r) *is recurrent.*

(2) $e \in \mathbf{D}_0$.

(3) $\mathbf{D} = \mathbf{D}_0$.

Proof. The equivalence of (1) and (2) is an immediate consequence of Corollary (3.42), since $D(u) = D(u - e)$.

Obviously (3) implies (2), so that we have only to show the (2) implies (3). Assume that (2) holds and let u_n be a sequence of finitely supported functions such that $\|u_n - e\| \to 0$. Let $u_n^+ = \max(u_n, 0)$. Then u_n^+ is a finitely supported function and $\|u_n^+ - e\| \le \|u_n - e\|$. Passing to a subsequence if necessary, we may assume that $\|u_n^+ - e\| \le n^{-2}$. Therefore the functions $v_n = nu_n^+$ are finitely supported non–negative functions. They also satisfy $D(v_n) \le n^{-2}$ and, for every $x \in V$, $v_n(x) \to \infty$ (since $u_n^+(x) \to 1$).

Suppose first that f is any non–negative function in $\mathbf{D}$ and set $f_n = \min(f, v_n)$. Then, f_n is finitely supported and $f_n(x) \to f(x)$ pointwise. In particular, $(f - f_n)(o) = 0$ for large values of n.

Set $Y_n = \{[x, y] \in Y : f(x) - f(y) \ne f_n(x) - f_n(y)\}$. For every $\epsilon > 0$ $\sum_{[x,y] \in Y_n} c(x, y)|f(x) - f(y)|^2 < \epsilon$ if n is sufficiently large. If $[x, y] \in Y_n$ and $f(x) > v_n(x)$, $f_n(y) > v_n(y)$, then

$$|f(x) - f(y) - f_n(x) + f_n(y)|^2 = |f(x) - f(y) - v_n(x) + v_n(y)|^2$$
$$\le 2|f(x) - f(y)|^2 + 2|v_n(x) - v_n(y)|^2.$$

If $[x, y] \in Y_n$ and, say, $f(x) \le v_n(x)$, $f(y) > v_n(y)$, then

$$|f(x) - f(y) - f_n(x) + f_n(y)|^2 = \left(f(y) - v_n(y)\right)^2$$
$$\le \left(f(y) + v_n(x) - f(x) - v_n(y)\right)^2$$
$$\le 2|f(x) - f(y)|^2 + 2|v_n(x) - v_n(y)|^2.$$

It follows that, for sufficiently large values of n,

$$(3.64) \quad D(f - f_n) = \frac{1}{2} \sum_{[x,y] \in Y_n} c(x, y)|f(x) - f(y) - f_n(x) + f_n(y)|^2$$
$$\le \sum_{[x,y] \in Y_n} c(x, y)|f(x) - f(y)|^2 + 2D(v_n) \le \epsilon + 2n^{-2}.$$

Therefore $f \in \mathbf{D}_0$.

Finally, given any function in $\mathbf{D}$, the above arguments show that its positive and negative parts are in $\mathbf{D}_0$. $\square$

DEFINITION. Let (Γ, r) be an infinite electrical network. We denote by $\mathbf{HD}$ the subspace of all harmonic functions u on V such that $D(u) < \infty$.

Let u and v be real valued Dirichlet finite functions. In the sequel we will use the notation

$$(3.65) \qquad \lfloor u, v \rfloor = \frac{1}{2} \sum_{x \sim y} c(x, y)(u(x) - u(y))(v(x) - v(y)).$$

We have the following Lemma.

Lemma (3.66). *The space* **HD** *is closed in* **D***. Furthermore,* $\lfloor u, v \rfloor = 0$ *for every* $u \in$ **HD** *and* $v \in \mathbf{D}_0$.

Proof. For every finitely supported w and every $u \in \mathbf{D}$

$$(3.67) \qquad \lfloor u, w \rfloor = \sum_{x \in V} c(x)w(x)(1 - P)(u)(x).$$

If u is harmonic, then $\lfloor u, w \rfloor = 0$.

Now, observe that

$$(3.68) \qquad \lfloor u, w \rfloor \le D(u)^{1/2} D(w)^{1/2}$$

for all $u, w \in \mathbf{D}$. For every $v \in \mathbf{D}_0$ there exists a sequence of functions $w_n \in \ell_0$ such that $D(v - w_n) \to 0$ as $n \to \infty$. Thus $\lfloor u, v \rfloor = 0$, by (3.68).

Next, suppose that $\{u_n\}$ is a sequence of function in **HD** which is norm convergent to u. Then, for every $x \in V$

$$\sum_{y \in V(x)} p(x,y)|u_n(y) - u(y)| \le \left(c(x)^{-1} \sum_{y \in V(x)} c(x,y)|u_n(y) - u(y)|^2 \right)^{1/2}$$

$$= c(x)^{-1/2} D(u - u_n)^{1/2}.$$

Hence $u_n(x) = P(u_n)(x) \to P(u)(x)$ for all $x \in V$. Since u_n is pointwise convergent to u, we have $u = P(u)$. $\square$

The following Theorem is the discrete version of Royden's decomposition theorem for Dirichlet finite functions on Riemannian surfaces (see Royden (1952)).

Theorem (3.69). *Suppose that* (Γ, r) *is a transient network. For every* $f \in \mathbf{D}$ *there exist a unique* $u \in$ **HD** *and a unique* $v \in \mathbf{D}_0$ *such that* $f = u + v$*. For such functions* u *and* v *we have* $D(f) = D(u) + D(v)$*.*

Proof. Let $\alpha = \inf_{w \in \mathbf{D}_0} D(f - w)$. The functional $w \mapsto D(f - w)$ is strictly convex and lower weakly semicontinuous. Therefore there exists a unique $v \in \mathbf{D}_0$ such that $\alpha = D(f - v)$. Set $u = f - v$. We show that u is harmonic.

For all finitely supported w and real t

$$D(u - tw) = D(g) + t^2 D(w) - 2t \lfloor u, w \rfloor$$

so that $D(u - tw) = D(g) + t^2 D(w)$ by Lemma (3.66). By the definition of u we have

$$0 = \lim_{t \to 0} \frac{D(u - tw) - D(u)}{t} = - \sum_x c(x)w(x)(1 - P(u))(x).$$

Taking $w = \delta_z$ for all $z \in V$, we see that u is harmonic. Moreover

$$D(u + v) = D(u) + D(v) + 2\lfloor u, v \rfloor = D(u) + D(v)$$

always by Lemma (3.66).

Finally, we prove the uniqueness of the decomposition. If $f = v' + u'$, where $v' \in \mathbf{D}_0$ and u' is harmonic, then $g = v - v' = u - u'$ is harmonic and belongs to $\mathbf{D}_0$. By Lemma (3.66) $D(g) = 0$ i.e., g is constant. Since the network is transient, and $g \in \mathbf{D}_0$, we get $g = 0$. □

The rest of this section is devoted to the study of harmonic and superharmonic functions in $\mathbf{D}$.

Lemma (3.70). *Suppose that (Γ, r) is a transient network and $u \in \mathbf{D}_0$. Then, the positive and negative parts of u belong to $\mathbf{D}_0$.*

If $u \in \mathbf{D}_0$ is superharmonic, then

(1) *$u(x) \geq 0$ for all x.*

(2) *there exists $f \geq 0$ such that $u = G(f)$.*

Proof. For any function h on V let $h^+ = \max(h, 0)$ and $h^- = -\min(h, 0)$. If $u_n \in \ell_0$ and $\|u - u_n\| \to 0$ as $n \to \infty$, then $\|u_n^+\| \leq \|u_n\|$ and $\|u_n^-\| \leq \|u_n\|$ are bounded.

Passing to a subsequence if necessary, u_n^+ and u_n^- converge weakly to functions $v^+ \in \mathbf{D}_0$ and $v^- \in \mathbf{D}_0$ respectively. By Lemma (3.14) $|u_n^+(x) - u^+(x)| \leq |u_n(x) - u(x)| \leq K_x \|u_n - u\|$ for all $x \in V$. It follows that $v^+ = u^+$ and, analogously, $v^- = u^-$. Hence u^+ and u^- are in $\mathbf{D}_0$, so that $|u| = u^+ + u^-$ belongs to $\mathbf{D}_0$.

Observe now that $D(|u|) \leq D(u)$. On the other hand, since $|u| = u + 2u^-$, we have

(3.71) $$D(|u|) = D(u) + 4D(u^-) + 4\lfloor u^-, u \rfloor.$$

Assume that u is superharmonic. Since $u_n^- \geq 0$ is finitely supported,

$$\sum_{x \sim y} c(x, y)(u_n^-(x) - u_n^-(y))(u(x) - u(y)) = 2 \sum_x c(x) u_n^-(x)(1 - P)(u)(x) \geq 0.$$

As u_n^- is weakly convergent to u^-, $\sum_{x \sim y} c(x, y)(u^-(x) - u^-(y))(u(x) - u(y)) \geq 0$. By (3.71) we get $D(u) \leq D(|u|)$. Hence $D(|u|) = D(u)$ and, always by (3.71), $D(u^-) = 0$. It follows that u^- is constant. By Theorem (3.63) $u^- = 0$. Thus (1) is proved.

To prove (2) we apply the Riesz decomposition theorem (1.38) to write u in the form $u = G(f) + v$, where f and v are non–negative. Choose any exhaustion Γ_n of of Γ and let $f_n(x) = f(x)$ if $x \in V_n$, $f_n(x) = 0$ otherwise. Set $u_n = G(f_n)$. Note that $u_n \in \mathbf{D}_0$ by Corollary (3.32). Remembering Lemma (1.32),

$$D(u_n) = (u_n, (1 - P)(u_n)) = \sum_x c(x)G(f_n)(x)f_n(x),$$

where $(\cdot, \cdot)$ denotes the inner product in ℓ^2. Hence, using Lemma (1.32) again, we have

$$D(u_n) \leq \sum_x c(x)G(f)(x)f_n(x) \leq \sum_x c(x)u(x)(1 - P)(u_n)(x)$$

$$= \frac{1}{2} \sum_x c(x, y)(u(x) - u(y))(u_n(x) - u_n(y)) \leq D(u)^{1/2} D(u_n)^{1/2}.$$

Thus $D(u_n)$ is bounded by $D(u)$. Since $u_n(o) \leq u(o)$, passing to a subsequence if necessary, we have that u_n is weakly convergent to a function $g \in \mathbf{D}_0$. On the other hand, by the dominated convergence theorem, u_n converges pointwise to $G(f)$. By Theorem (3.15), $g = G(f)$. Finally, by the uniqueness of Royden's decomposition, $v = 0$ and $u = G(f)$. $\square$

REMARK. As a consequence of (2) of Lemma (3.70) we see that, in the Royden decomposition Theorem (3.69), if f is superharmonic, then $v \geq 0$.

Theorem (3.72). *Suppose that $u \in$HD. Then, there are nonnegative harmonic functions u_1 and u_2 in* **HD** *such that $u = u_1 - u_2$.*

Proof. If the network is recurrent, then u must be constant and the thesis is obvious. Suppose that the network is transient.

Let $f_1 = \max(u, 0)$ and $f_2 = \max(-u, 0)$. Since f_1 and f_2 have finite Dirichlet sum, by Royden's decomposition theorem there are $v_j \in \mathbf{D}_0$ and $u_j \in \mathbf{HD}$ $(j = 1, 2)$ such that $-f_j = v_j + u_j$. Since $-f_j$ is superharmonic, v_j must be superharmonic too. By Lemma (3.70) v_j is nonnegative. It follows that $-u_j$ is nonnegative.

From the relation $u = f_1 - f_2 = (v_2 - v_1) + (u_2 - u_1)$ and the uniqueness of Royden's decomposition, we conclude that $v_1 = v_2$ and $u = u_2 - u_1$. $\square$

The following result is the discrete version of Virtanen's Theorem.

Theorem (3.73). *For every $h \in$ **HD** there exists a sequence h_n of bounded functions in* **HD** *such that $\|h - h_n\| \to 0$ as $n \to \infty$. If $h \geq 0$, then $0 \leq h_n \leq h$.*

Proof. We may assume that the network is transient. We may suppose that h is non–negative. Let, for all $n > 0$, $g_n = \min(h, n)$. Obviously g_n is superharmonic.

By Royden's Decomposition Theorem there exist functions s_n in $\mathbf{D}_0$ and h_n in **HD** such that $g_n = s_n + h_n$. Furthermore, s_n is superharmonic so that, by

Lemma (3.70), s_n is non–negative and $s_n = G(f_n)$ for some non–negative f_n. By the uniqueness of the Riesz decomposition (Theorem (1.38)), $h_n \geq 0$ and h_n is the greatest harmonic minorant of g_n. Therefore $0 \leq h_n \leq g_n \leq h$. We have

$$(3.74) \qquad D(h - g_n) = D(s_n) + D(h - h_n) - 2\lfloor h - h_n, s_n \rfloor.$$

Since $s_n \in \mathbf{D}_0$ and $h - h_n$ is harmonic, the last summand in (3.74) is zero by Lemma (3.66). Thus $D(h - g_n) = D(s_n) + D(h - h_n)$. As in the proof of inequalities (3.64), denote by Y_n be the set of edges $[x, y]$ such that $h(x) - h(y) \neq g_n(x) - g_n(y)$. Then, for every $\epsilon > 0$, there is an n_0 such that for all $n > n_0$

$$D(h - g_n) = \frac{1}{2} \sum_{[x,y] \in Y_n} c(x,y)|h(x) - h(y) - g_n(x) + g_n(y)|^2$$

$$\leq \frac{1}{2} \sum_{[x,y] \in Y_n} c(x,y)|h(x) - h(y)|^2$$

$$< \epsilon.$$

Thus $D(h_n - h) \to 0$ and $D(s_n) \to 0$. Now $0 \leq s_n \leq h_n \leq h$. Hence (passing to a subsequence if necessary) $s_n(o)$ converges to some constant κ. Hence s_n converges in norm to the constant function κ. Since the network is transient, $\kappa = 0$ by Theorem (3.63). So $h_n(o) \to h(o)$. This concludes the proof. $\square$

§8. Extremal length

The notion of extremal length of a set of paths in a network is the discrete counterpart of the notion of extremal length of a family of curves in a Riemannian manifold. It was first introduced by Duffin (1962) for finite networks and was subsequently studied in the infinite case by Yamasaki, Nakamura and Yamasaki, Kayano and Yamasaki. We start by introducing the extremal length in a network and by establishing its basic properties.

Let (Γ, r) be a network and let $\mathbf{P}$ be a family of (finite or infinite) paths of Γ. For every $\mathbf{p} \in \mathbf{P}$ let $Y(\mathbf{p})$ denote the edge set of $\mathbf{p}$. Denote by $Q(\mathbf{P})$ the subset of all 1–chains $I = \{i(x,y)\}$ such that, for all $\mathbf{p} \in \mathbf{P}$,

$$\mathcal{W}(I) < \infty \quad \text{and} \quad \frac{1}{2} \sum_{[x,y] \in Y(\mathbf{p})} r(x,y)|i(x,y)| \geq 1.$$

DEFINITION. The extremal length $\lambda(\mathbf{P})$ of $\mathbf{P}$ is defined by the following formula

$$\lambda(\mathbf{P})^{-1} = \inf\{\mathcal{W}(I) : I \in Q(\mathbf{P})\}.$$

Lemma (3.75). *The extremal length has the following properties:*

(1) *If* $\mathbf{P}_1 \subseteq \mathbf{P}_2$, *then* $\lambda(\mathbf{P}_1) \geq \lambda(\mathbf{P}_2)$;

(2) *If* $\{\mathbf{P}_n\}$ *is a countable family of paths and* $\mathbf{P} = \bigcup_{n=1}^{\infty} \mathbf{P}_n$, *then* $\lambda(\mathbf{P})^{-1} \leq \sum_{n=1}^{\infty} \lambda(\mathbf{P}_n)^{-1}$;

(3) $\lambda(\mathbf{P}) = \infty$ *for a set of paths* $\mathbf{P}$ *if and only if there exists* $I \in \mathbf{H}_1$ *such that* $\sum_{B \in Y(\mathbf{p})} r(B)|i(B)| = \infty$ *for all paths* $\mathbf{p} \in \mathbf{P}$.

Proof. The first assertion is an immediate consequence of the definition. To prove (2) we may assume that $\sum_{n=1}^{\infty} \lambda(\mathbf{P}_n)^{-1} < \infty$. For every $\epsilon > 0$ and $n > 0$ there exists a 1–chain I_n such that $\mathcal{W}(I_n) < \lambda(\mathbf{P}_n)^{-1} + 2^{-n}\epsilon$. Choose any orientation X of Γ and define a 1–chain I on Γ by letting $i(x,y) = \sup_n |i_n(x,y)|$, where $i_n(x,y)$ is the value of I_n on $[x,y] \in X$. Note that $i(x,y)$ is finite since $\lambda^{-1}(\mathbf{P}_n)$ is bounded. We have

$$(3.76) \qquad \epsilon + \sum_{n=1}^{\infty} \lambda(\mathbf{P}_n)^{-1} \geq \sum_{n=1}^{\infty} \mathcal{W}(I_n) \geq \mathcal{W}(I) \geq \lambda(\mathbf{P})^{-1},$$

so that (2) holds.

Suppose now that there exists a 1–chain $I \in \mathbf{H}_1$ such that

$$\sum_{[x,y] \in Y(\mathbf{p})} r(x,y)|i(x,y)| = \infty \quad \text{for all } \mathbf{p} \in \mathbf{P}.$$

For every $\epsilon > 0$ we have $\lambda(\mathbf{P})^{-1} \leq \mathcal{W}(\epsilon I) = \epsilon^2 \mathcal{W}(I)$. Letting ϵ tend to 0 we obtain $\lambda(\mathbf{P}) = \infty$.

Conversely let $\lambda(\mathbf{P}) = \infty$. Then, there is a sequence $I_n \in \mathbf{H}_1$ such that $\mathcal{W}(I_n) < 4^{-n}$ and such that $\frac{1}{2} \sum_{Y(\mathbf{p})} r(x,y)|i_n(x,y)| \geq 1$. Choose any orientation X of Γ and define a 1–chain I by setting

$$i(x,y) = \Big(\sum_{n=1}^{\infty} 2^n i_n^2(x,y)\Big)^{1/2} \quad \text{for all } [x,y] \in X.$$

Clearly $I \in \mathbf{H}_1$. Moreover, for every $n > 0$ and every $\mathbf{p} \in \mathbf{P}$,

$$\sum_{[x,y] \in Y(\mathbf{p})} r(x,y)|i(x,y)| = \sum_{[x,y] \in Y(\mathbf{p})} r(x,y)\Big(\sum_{k=1}^{\infty} 2^k i_k^2(x,y)\Big)^{1/2}$$

$$(3.77) \qquad\qquad\qquad \geq 2^{n/2} \sum_{[x,y] \in Y(\mathbf{p})} r(x,y)|i_n(x,y)|$$

$$\geq 2 \cdot 2^{n/2}.$$

Letting n tend to ∞ in (3.77) we obtain (3). $\square$

DEFINITION. Let $a \in V$ be a vertex of Γ. We denote by $\mathbf{P}_a$ the set of all infinite one–sided paths $\mathbf{p}$ of Γ whose first vertex is a. If b is another vertex of Γ we denote by $\mathbf{P}_{a,b}$ the set of all finite paths joining a and b.

Our next aim is to show that, for locally finite networks, the extremal length of $\mathbf{P}_a$ coincides with the resistance between a and infinity. In order to establish this result, we first prove the analogous result for finite networks. The following Theorem was proved by Duffin (1962).

Theorem (3.78). *Suppose that (Γ, r) is a finite network. Let a and b be distinct vertices and let R denote the resistance between a and b. Then, $R = \lambda(\mathbf{P}_{a,b})$.*

Proof. For every path $\mathbf{p} \in \mathbf{P}_{a,b}$ let $a = x_0 \sim x_1 \sim \cdots \sim x_n = b$ denote the vertices of $\mathbf{p}$. Let u be a function on V such that $u(a) = 1$, $u(b) = 0$ and set $I = du = \{c(x,y)(u(x) - u(y))\}$. Then I is a 1–chain such that $\mathcal{W}(I) = D(u)$ and

$$\frac{1}{2} \sum_{[x,y] \in \Upsilon(\mathbf{p})} r(x,y)|i(x,y)| \geq \sum_{j=1}^{n} (u(x_{j-1}) - u(x_j)) = 1.$$

Therefore $R \leq \lambda(\mathbf{P}_{a,b})$ by the Dirichlet principle for finite networks (Theorem (2.16)).

To prove the converse inequality, for every $x \neq a$ we define a walk $\mathbf{W}$ from a to x to be a finite sequence of not necessarily distinct vertices $x_0, \ldots, x_n$ in V such that $x_0 = a$, $x_n = x$ and $x_{j-1} \sim x_j$, $j = 1, \ldots, n$.

For every $\epsilon > 0$ there is a 1–chain $I \in Q(\mathbf{P}_{a,b})$ such that $\mathcal{W}(I) < \lambda(\mathbf{P}_{a,b})^{-1} + \epsilon$. We define a function u on V by setting $u(a) = 0$ and, for $x \neq a$,

$$u(x) = \inf_{\mathbf{W}} \sum_{j=1}^{n} r(x_{j-1}, x_j)|i(x_{j-1}, x_j)|,$$

where the infimum is taken over all walks $\mathbf{W}$ from a to x.

Let $x \sim y$. For every walk $\mathbf{W}$ from a to x, say $x_0 = a \sim x_1 \sim \cdots \sim x_n = x$, we define a walk $\mathbf{W}'$ from a to y simply by adding the vertex y to $\mathbf{W}$. Therefore, for every $\mathbf{W}$

$$u(y) \leq \sum_{j=1}^{n} r(x_{j-1}, x_j)|i(x_{j-1}, x_j)| + r(x,y)|i_n(x,y)|.$$

Taking the infimum over all walks from a to x we get $u(y) \leq u(x) + r(x,y)|i(x,y)|$. Reversing the role of x and y we obtain also $u(x) \leq u(y) + r(x,y)|i(x,y)|$ i.e., $|u(x) - u(y)| \leq r(x,y)|i(x,y)|$. Therefore $D(u) \leq \mathcal{W}(I) < \lambda(\mathbf{P}_{a,b})^{-1} + \epsilon$. Setting

$w = \max(1 - u, 0)$ we obtain that $w(a) = 1$, $w(b) = 0$ and $D(w) \leq D(u)$. Hence $R \geq \lambda(\mathbf{P}_{a,b})$ by Dirichlet's principle again. $\square$

Suppose that (Γ_n, r_n) is an exhaustion of a locally finite network (Γ, r), and let $a \in V_1$. Let (Γ'_n, r'_n) be as in Theorem (3.25). For every 1–chain I on Γ let $I^n = \{i^n(x, y)\}$ denote the restriction of I to Γ'_n. We set

$$t(I) = \inf\{\frac{1}{2} \sum_{[x,y] \in Y(\mathbf{p})} r(x,y)|i(x,y)| : \mathbf{p} \in \mathbf{P}_a\}$$

$$t_n(I) = \inf\{\frac{1}{2} \sum_{[x,y] \in Y(\mathbf{p})} r'_n(x,y)|i^n(x,y)| : \mathbf{p} \in \mathbf{P}_{a,b_n}\}.$$

Lemma (3.79). *If (Γ, r) is a locally finite network, then*

$$\lim_{n \to \infty} t_n(I) = t(I).$$

Proof. By local finiteness we may assume that a is not a boundary point of Γ_n. Let ϕ_n be the morphism of the definition of Γ'_n. By the definition of I^n and r'_n

$$r'_n(B')|i^n(B')| = r'_n(B')|\sum_{\{B: \Phi_n(B) = B'\}} i(B)| \leq \sum_{\{B: \Phi_n(B) = B'\}} r(B)|i(B)|.$$

Hence the sequence $\{t_n(I)\}$ is bounded from above by $t(I)$. The same argument shows that $t_n(I) \leq t_{n+1}(I)$.

For each n there exists a path $\mathbf{p}_n \in \mathbf{P}_{a,b_n}$ such that

$$t_n(I) = \frac{1}{2} \sum_{B' \in Y(\mathbf{p}_n)} r'_n(B')|i^n(B')|.$$

Since Γ is locally finite, there exists $x_1 \sim a$ such that x_1 is a vertex of $\mathbf{p}_n$ for infinitely many n. By induction, there is a sequence of vertices $x_j \in V$ such that

$$a = x_0 \sim x \sim x_2 \sim \cdots \sim x_n \sim \cdots.$$

and such that for every m there are infinitely many $\mathbf{p}_n$ whose first $m + 1$ vertices are $x_0, \ldots, x_m$. For every $\epsilon > 0$ we have for n large enough

$$(3.80) \qquad \frac{1}{2} \sum_{j=0}^{m} r(x_{j-1}, x_j)|i(x_{j-1}, x_j)| \leq \frac{1}{2} \sum_{B' \in Y(\mathbf{p}_n)} r'(B')|i^n(B')| = t_n(I)$$

$$\leq \lim_{n \to \infty} t_n(I) + \epsilon.$$

Letting m tend to ∞ in (3.80) we obtain

$$t(I) \leq \frac{1}{2} \sum_{j=0}^{m} r(x_{j-1}, x_n)|i(x_{j-1}, x_n)| \leq \lim_{n \to \infty} t_n(I) + \epsilon.$$

Theorem (3.81). *Let (Γ, r) be a locally finite infinite network, and let $a \in V$. Let $R(a)$ denote the effective resistance between a and ∞. Then*

$$R(a) = \lambda(\mathbf{P}_a).$$

Proof. Arguing as in the first part of the proof of Theorem (3.78), and using Dirichlet principle for infinite networks (Theorem (3.41)), we obtain $R(a) \leq \lambda(\mathbf{P}_a)$. Therefore the theorem is true for recurrent networks. Thus let us assume that (Γ, r) is transient.

Let (Γ_n, r_n) be any exhaustion of Γ and let (Γ'_n, r'_n) be as in Theorem (3.25). Let $R'_n = c(a)^{-1} G(a, a; b_n)$ be the resistance between a and b_n. Then, denoting by λ'_n the extremal length in (Γ'_n, r'_n), $\lambda'_n(\mathbf{P}_{a,b_n}) = R'_n \to R(a)$.

For every $\epsilon > 0$ let $I \in Q(\mathbf{P}_a)$ be such that

$$\mathcal{W}(I) \leq \lambda(\mathbf{P}_a)^{-1} + \epsilon.$$

By Lemma (3.79) we have, as $n \to \infty$,

(3.82) $$\mathcal{W}\left(\frac{I}{t_n(I)}\right) \to \mathcal{W}\left(\frac{I}{t(I)}\right) \leq \mathcal{W}(I) \leq \lambda(\mathbf{P}_a)^{-1} + \epsilon.$$

Denote by I^n the restriction of I to Γ'_n. Then (as $I^n/t_n(I) \in Q(\mathbf{P}_{a,b_n})$),

(3.83) $$\mathcal{W}\left(\frac{I}{t_n(I)}\right) \geq \mathcal{W}\left(\frac{I^n}{t_n(I)}\right) \geq \lambda'_n(\mathbf{P}_{a,b_n})^{-1}.$$

Combining (3.82) and (3.83) we have $\lambda(\mathbf{P}_a)^{-1} \geq \lambda'_n(\mathbf{P}_{a,b_n})^{-1} - 2\epsilon$ for n large enough. Therefore $R^{-1}(a) \leq \lambda(\mathbf{P}_a)^{-1}$. $\square$

Corollary (3.84). *An infinite locally finite network is recurrent if and only if there exists a vertex a such that $\lambda(\mathbf{P}_a) = \infty$. In this case $\lambda(\mathbf{P}_x) = \infty$ for every vertex x.*

§9. Limits of Dirichlet functions along paths

DEFINITION. Suppose that $\mathbf{P}$ is a set of paths. We will say that a property holds for almost every path in $\mathbf{P}$ if the subset of all paths for which the property is not true has extremal length ∞.

Theorem (3.85). *Let (Γ, r) be an infinite network and let $u \in \mathbf{D}$. Then, for almost every one-sided infinite path $\mathbf{p}$ in Γ, $u(x)$ converges as x tends to ∞ along the vertices of $\mathbf{p}$.*

Proof. Let $a \in V$ and let $\mathbf{p}$ be a one-sided infinite path in $\mathbf{P}_a$ with vertices $x_0 = a \sim x_1 \sim \cdots$. Let $\mathbf{P}$ be the subset of all paths in $\mathbf{P}_a$ such that $\sum_{j=1}^{\infty} |u(x_{j-1}) -$

$u(x_j)| = \infty$. Then, for all $\mathbf{p} \in \mathbf{P}_a$ not in $\mathbf{P}$ we have that $\sum_{j=1}^{n}(u(x_j - x_{j-1})) = -u(a) + u(x_n)$ converges as n tends to ∞.

The 1–chain $I = \{i(x, y)\}$ such that $i(x, y) = c(x, y)(u(x) - u(y))$ satisfies $\mathcal{W}(I) = D(u) < \infty$ and $\sum_{[x,y] \in Y(\mathbf{p})} r(x, y)|i(x, y)| = \infty$ for all $\mathbf{p} \in \mathbf{P}$. Hence $\lambda(\mathbf{P}) = \infty$ by (3) of Lemma (3.75). Since Γ is countable, the thesis now follows from (2) of Lemma (3.75) $\square$

A function $v \in \mathbf{D}_0$ does not necessarily vanish at infinity (e.g. a nonzero constant function on a recurrent network). However $u(x)$ must tend to zero along the vertices of almost all paths. See Glasner and Katz (1982) for the case of Riemannian manifolds.

Theorem (3.86). *Let (Γ, r) be an infinite network and let $v \in \mathbf{D}_0$. Then, for every $a \in V$ and for almost every one–sided infinite path $\mathbf{p} \in \mathbf{P}_a$ of Γ, $\lim v(x) = 0$ as x tends to ∞ along the vertices of $\mathbf{p}$.*

Proof. Let v_k be a sequence of finitely supported functions converging to v in $\mathbf{D}_0$. Define 1–chains I_k and I by $I_k = dv_k$ (i.e., $i_k(x, y) = c(x, y)(v_k(x) - v_k(y))$ and $I = dv$ (i.e., $i(x, y) = c(x, y)(v(x) - v(y))$. Then $\mathcal{W}(I - I_k) \to 0$ as $k \to \infty$.

Let $a \in V$ and denote by $\mathbf{P}^0$ the subset of all paths in $\mathbf{P}_a$ with the property that $\sum_{[x,y] \in Y(\mathbf{p})} r(x, y)|i(x, y)| < \infty$. Then, as in the proof of Theorem (3.85), we see that $\lambda(\mathbf{P}_a \setminus \mathbf{P}^0) = \infty$. Therefore

$$(3.87) \qquad \lim_{n \to \infty} v(x_n) = v(a) + \sum_{j=0}^{\infty}(v(x_j) - v(x_{j-1}))$$

exists as x tends to infinity along the vertices $x_0 \sim x_1 \sim \cdots$ of every $\mathbf{p} \in \mathbf{P}^0$. We denote by $v(\mathbf{p})$ such a limit.

Let (Γ_n, r_n) be an exhaustion of (Γ, r), $\Gamma_n = (V_n, Y_n)$, $a \in V_1$. For every $\epsilon > 0$ set

$$\mathbf{P}(\epsilon) = \{\mathbf{p} \in \mathbf{P}^0 : |v(\mathbf{p})| \geq \epsilon\}.$$

For every n and every path $\mathbf{p}$ with vertices $a = x_0 \sim x_1 \sim \cdots$, let $X(\mathbf{p})$ denote the set of all oriented edges $[x_{j-1}, x_j]$ of $\mathbf{p}$ and set $X_n(\mathbf{p}) = X(\mathbf{p}) \cap Y_n$. Define

$$\mathbf{P}^n(\epsilon) = \{\mathbf{p} \in \mathbf{P}(\epsilon) : | \sum_{[x,y] \in X_m(\mathbf{p})} r(x, y)i(x, y) + v(a) - v(\mathbf{p})| < \frac{\epsilon}{4} \text{ for all } m \geq n\}.$$

Since Y_m is a finite set, for all m there exists $k_1 = k_1(m)$ such that for all $k > k_1$

$$(3.88) \qquad \begin{aligned} &|v(a) - v_k(a)| < \frac{\epsilon}{4} \\ &\sum_{[x,y] \in Y_m} r(x, y)|i(x, y) - i_k(x, y)| < \frac{\epsilon}{4}. \end{aligned}$$

Note that k_1 depends on m but does not depend on any $\mathbf{p}$. Let $n < m$ be fixed and let $\mathbf{p} \in \mathbf{P}^n(\epsilon)$. For $k > k_1$ we have by (3.88)

$$\sum_{[x,y] \in X(\mathbf{p}) \setminus X_m(\mathbf{p})} r(x,y)|i_k(x,y)|$$

$$\geq \Big| \sum_{[x,y] \in X(\mathbf{p})} r(x,y)i_k(x,y) - \sum_{[x,y] \in X_m(\mathbf{p})} r(x,y)i_k(x,y) \Big|$$

$$(3.89) \qquad = \Big| \sum_{[x,y] \in X_m(\mathbf{p})} r(x,y)i_k(x,y) + v_k(a) \Big|$$

$$> \Big| \sum_{[x,y] \in X_m(\mathbf{p})} r(x,y)i(x,y) + v(a) \Big| - \frac{\epsilon}{2}$$

$$> |v(\mathbf{p})| - \frac{3\epsilon}{4} > \frac{\epsilon}{4}.$$

Let X be an orientation of Γ and let $X_m = Y_m \bigcap X$ be the induced orientation of Γ_m. For every $k > k_1$ let $J_k = \{j_k(x,y)\}$ be the 1–chain such that $j_k(x,y) = 4\epsilon^{-1}|i_k(x,y)|$ if $[x,y] \in X \setminus X_m$ and 0 otherwise.

By (3.89)

$$\sum_{[x,y] \in X} r(x,y)|j_k(x,y)| \geq 1 \quad \text{for all } \mathbf{p} \in \mathbf{P}^n(\epsilon)$$

and

$$\lambda(\mathbf{P}^n(\epsilon))^{-1} \leq \mathcal{W}(J_k) \leq \frac{16}{\epsilon^2} \sum_{[x,y] \in X \setminus X_m} r(x,y)|i_k(x,y)|^2.$$

Letting k tend to ∞ we get for every $m > n$

$$(3.90) \qquad \lambda(\mathbf{P}^n(\epsilon))^{-1} \leq \frac{16}{\epsilon^2} \sum_{[x,y] \in X \setminus X_m} r(x,y)|i(x,y)|^2.$$

Letting m tend to ∞ in (3.90) we obtain that $\lambda(\mathbf{P}^n(\epsilon))^{-1} = 0$ for all n. Since $\mathbf{P}(\epsilon) = \bigcup_{n=1}^{\infty} \mathbf{P}^n(\epsilon)$, we have $\lambda(\mathbf{P}(\epsilon)) = \infty$ by (2) of Lemma (3.75).

Let $\mathbf{P}^* = \{\mathbf{p} \in \mathbf{P}^0 : v(\mathbf{p}) \neq 0\}$. We have $\mathbf{P}^* = \bigcup_{n=1}^{\infty} \mathbf{P}(1/n)$. Therefore, always by (2) of lemma (3.75), $\lambda((\mathbf{P}_a \setminus \mathbf{P}^0) \cup \mathbf{P}^*) = \infty$. Hence $v(\mathbf{p}) = 0$ for almost every $\mathbf{p} \in \mathbf{P}_a$. $\square$

REMARK. Since Γ is countable $v(\mathbf{p}) = 0$ for every $v \in \mathbf{D}_0$ and for almost all one–sided infinite paths $\mathbf{p}$ in Γ.

Theorem (3.91). *Suppose that (Γ, r) is a transient locally finite network. Let $u \in \mathbf{HD}$. If there exists a constant c such that $\lim u(x) = c$ along the vertices of almost every one–sided infinite path $\mathbf{p}$, then $u(x) = c$ for all $x \in V$.*

Proof. For every $x \in V$ we set

$$V^+(x) = \{y \in V(x) : u(y) > u(x)\}.$$

Suppose that u is not constant. Then, there exists $a \in V$ such that there exists $x \sim a$ with the property that $u(x) \neq u(a)$. Let $U_0 = \{a\}$. Since u is harmonic, $V^+(a) \neq \emptyset$. We set $U_1 = U_0 \cup V^+(a)$. Suppose that $U_1, U_2, \ldots, U_{n-1}$ have been defined. We define U_n to be the set

$$(3.92) \qquad U_n = \bigcup_{x \in U_{n-1}} V^+(x).$$

Denote by Γ^+ the subgraph of Γ whose vertex set is $V^+ = \bigcup_{n=0}^\infty U_n(x)$ and whose edge set Y^+ is defined as follows. For every $x, y \in V^+$ such that $x \sim y$, the edge $[x, y]$ of Γ is also in Y^+ if and only if either $y \in V^+(x)$ or $x \in V^+(y)$. Of course we may fix an orientation X^+ of Γ^+ by choosing the edge $[x, y]$ whenever $u(y) > u(x)$.

The graph Γ^+ is connected, since by (3.92) any element of V^+ can be joined with a by means of a path in Γ^+. Furthermore, Γ^+ is infinite. To see this, it suffices to show that if U_{n-1} is finite, then $U_n \neq U_{n-1}$, $n \geq 2$. If U_{n-1} is finite, then there exists $z \in U_{n-1}$ such that $u(z)$ is maximum in U_{n-1}. There exists $x \in U_{n-2}$, $x \sim z$, such that $u(z) > u(x)$, so that $V^+(z) \neq \emptyset$. If $y \in V^+(z)$, then $u(y) > u(z)$ so that, by the definition of z, $y \in U_n \setminus U_{n-1}$.

Let r^+ denote the restriction of r to the edges of Γ^+. We claim that the subnetwork (Γ^+, r^+) is transient. Let $x \in V^+$ and denote by $N(x)$ the set of all neighbors of x in Γ^+. We set

$$(3.93) \qquad \begin{aligned} V^-(x) &= \{y \in V(x) : y \notin V^+\} \\ N^-(x) &= \{y \in N(x) : u(y) < u(x)\}. \end{aligned}$$

Clearly $V(x) = V^-(x) \cup N^-(x) \cup V^+(x)$ and the three sets are mutually disjoint. Moreover $N(x) = N^-(x) \cup V^+(x)$. Since $u(y) \leq u(x)$ for every $y \in V^-(x)$, we get

$$\begin{aligned} 0 &= \sum_{y \in V(x)} c(x, y)\big(u(x) - u(y)\big) \\ &\geq \sum_{y \in N^-(x)} c(x, y)\big(u(x) - u(y)\big) + \sum_{y \in V^+(x)} c(x, y)\big(u(x) - u(y)\big) \\ &= \sum_{y \in N(x)} c^+(x, y)\big(u(x) - u(y)\big). \end{aligned}$$

Therefore u is superharmonic in (Γ^+, r^+). Since, by Theorem (3.34), a recurrent network has no nonconstant superharmonic functions in $\mathbf{D}$, we conclude that (Γ^+, r^+) is transient.

If $\mathbf{P}_a^+ \subseteq \mathbf{P}_a$ denotes the set of all one–sided infinite paths $\mathbf{p}$ in Γ^+ which start at a, and if λ^+ denotes the extremal length on (Γ^+, r^+), then Corollary (3.84) implies

$$\lambda(\mathbf{P}_a^+) = \lambda^+(\mathbf{P}_a^+) < \infty.$$

Let $u(\mathbf{p})$ denote the limit of $u(x)$ as $x \to \infty$ along the vertices of a one-sided path $\mathbf{p}$. For every $\mathbf{p} \in \mathbf{P}_a^+$, $u(\mathbf{p}) > u(a)$.

Since u is harmonic and there exists $y \sim a$ such that $u(y) > u(a)$, there exists $z \sim a$ such that $-u(z) > -u(a)$. Repeating the above arguments for the harmonic function $-u$, we find another subset of paths $\mathbf{P}_a^- \subseteq \mathbf{P}_a$ such that $\lambda(\mathbf{P}_a^-) < \infty$ and $u(\mathbf{p}) < u(a)$ for all $\mathbf{p} \in \mathbf{P}_a^-$.

Therefore, unless u is a constant function, there is no constant c such that $u(\mathbf{p}) = c$ for almost every path $\mathbf{p}$. $\square$

Thus we get the following characterization of the functions in $\mathbf{D}_0$.

Corollary (3.94). *Let (Γ, r) be a locally finite network. A function $f \in \mathbf{D}$ belongs to $\mathbf{D}_0$ if and only if $\lim f(x) = 0$ as $x \to \infty$ along the vertices of almost every one-sided infinite path.*

Proof. If the network is recurrent, then $\mathbf{D} = \mathbf{D}_0$ and the thesis is obvious by Corollary (3.84). If the network is transient, then we apply Royden's Decomposition Theorem and write f as a sum of a function v in $\mathbf{D}_0$ and of a harmonic function $u \in \mathbf{HD}$. The assertion now follows from Theorems (3.91) and (3.86). $\square$

We remark also also the following consequence.

Corollary (3.95). *Let (Γ, r) and u be as in Theorem (3.91). If there exists $o \in V$ and a constant c such that $\lim u(x) = c$ along the vertices of almost every one-sided infinite path $\mathbf{p} \in \mathbf{P}_o$, then $u(x) = c$ for all $x \in V$.*

Proof. This Corollary follows immediately from Theorem (3.91) and the connectedness of Γ. $\square$

CHAPTER IV

UNIQUENESS AND RELATED TOPICS

§1. Spaces of cycles

Let (Γ, r) be an electrical network and let $\imath$ be a 0–chain in $\partial \mathbf{H}_1$. We have seen (Chapter III, Theorem (3.13)) that solving Kirchhoff's equations (3.6), (3.7) in $\mathbf{H}_1$ is equivalent to solving in $\mathbf{D}$ the discrete Poisson's equation

$$(4.1) \qquad u(x) - \sum_{y \in V} p(x, y) u(y) = f(x) \qquad \text{for all } x \in V,$$

where $f(x) = -c(x)^{-1} \imath(x)$. In particular, $\imath$ generates a unique current in $\mathbf{H}_1$ if and only if there are no nonconstant harmonic functions in $\mathbf{D}$. Therefore, we will study the uniqueness of currents in $\mathbf{H}_1$ by investigating necessary and/or sufficient conditions for the space $\mathbf{HD}$ to be trivial.

We start by proving a simple condition equivalent to uniqueness, indicated by Doyle in his manuscript in (1988). This condition is stated in terms of the minimal and the limit current (see Chapter III, §2 and following).

Let $\mathbf{Z}^*$ denote the subspace of all cycles in $\mathbf{H}_1$ and let $\mathbf{Z}$ denote the closure in $\mathbf{H}_1$ of the subspace of all finite cycles.

Lemma (4.2). *The homogeneous Kirchhoff's equations*

$$(4.3) \qquad \begin{aligned} &\partial I = 0, \\ &(I, Z) = 0 \quad \text{for all finite cycles } Z \end{aligned}$$

have only the trivial solution in $\mathbf{H}_1$ *if and only if* $\mathbf{Z} = \mathbf{Z}^*$.

Proof. Assume that $\mathbf{Z} \neq \mathbf{Z}^*$. Then there is a cycle in $\mathbf{Z}^*$ $I \in H_1$, not identically zero, which is orthogonal to $\mathbf{Z}$. It is clear that such a 1–chain satisfies equations (4.3).

Conversely, if there exists a nontrivial $I \in \mathbf{Z}^*$ satisfying the homogeneous Kirchhoff's equations, then $\mathbf{Z} \neq \mathbf{Z}^*$, by the second of (4.3). $\square$

Theorem (4.4). *A necessary and sufficient condition in order that equations (4.3) have only the trivial solution $I = 0$ in $\mathbf{H}_1$ is that, for every $a \sim b$ $(a \neq b)$, the minimal and the limit currents generated by the dipole $\imath = \delta_b - \delta_a$ coincide.*

Proof. We have to prove only sufficiency. Let us proceed by contradiction.

Taking Lemma (4.2) into account, assume that $\mathbf{Z} \neq \mathbf{Z}^*$. Then, there exists $\overline{Z} \in \mathbf{Z}^*$ such that $(Z, \overline{Z}) = 0$ for every finite cycle Z. Let $a \sim b$ be vertices such that $\overline{z}(a, b) \neq 0$ and denote by E the 1–chain which takes value 1 on $[a, b]$, -1 on $[b, a]$ and 0 elsewhere. Clearly $\partial E = \delta_a - \delta_b$.

Let I_M and I_L denote the minimal and the limit current, respectively, generated by $\imath = \delta_b - \delta_a$. By definition of minimal current (cfr. Theorem (3.21)), $I_M + E$ is the projection of E onto $\mathbf{Z}^*$. On the other hand, an inspection of the proof of Theorem (3.16) (or Theorem (3.21) again) shows that $I_L + E$ is the projection of the 1–chain E onto $\mathbf{Z}$.

By the definition of $\overline{Z}$ we have

$$(4.5) \qquad \begin{aligned} (I_M, \overline{Z}) &= 0 \\ (I_L, \overline{Z}) &= -(E, \overline{Z}). \end{aligned}$$

Clearly, $(E, \overline{Z}) = 2\overline{z}(a, b) \neq 0$, so that $I_L \neq I_M$ by (4.5). $\square$

Because of the similarities already remarked between the theory of Dirichlet spaces on a network and on a Riemannian manifold, we borrow the following notation from the classification theory of Riemannian manifolds.

DEFINITION. We say that a network (Γ, r) belongs to the class $\mathcal{O}_{\mathbf{HD}}$ if there are no noncostant harmonic functions on (Γ, r) with finite Dirichlet sum.

Theorem (3.34) implies the following result.

Theorem (4.6). *If (Γ, r) is recurrent then $(\Gamma, r) \in \mathcal{O}_{\mathbf{HD}}$.*

If a network is transient, then it belongs to $\mathcal{O}_{\mathbf{HD}}$ if and only if every Dirichlet finite function u on V has the Royden decomposition $u = f + \kappa$, where $f \in \mathbf{D}_0$ and κ is constant. This simple observation allows us to state a comparison theorem which may be useful in the applications.

Theorem (4.7). *Suppose that r and r' are resistances on the same graph Γ and assume that there is a constant k such that*

$$k^{-1} r(x, y) \leq r'(x, y) \leq k r(x, y) \quad \text{for all } x \sim y.$$

Then $(\Gamma, r) \in \mathcal{O}_{\mathbf{HD}}$ if and only if $(\Gamma, r') \in \mathcal{O}_{\mathbf{HD}}$.

Proof. For every f, $k^{-1} D(f) \leq D'(f) \leq k D(f)$, where $D'(f)$ is the Dirichlet sum of f with conductances $c'(x, y) = r'(x, y)^{-1}$. Hence the two spaces $\mathbf{D}$ and $\mathbf{D}'$

are isomorphic, as well as $\mathbf{D}_0$ and $\mathbf{D}_0'$ (with obvious meaning of the notation). This implies that the Royden decomposition is the same in the two networks, whence the thesis. $\square$

§2. Bounded automorphisms

DEFINITION. Let (Γ, r) be a network and let ϕ be an automorphism of Γ. We say that ϕ induces (or is) an automorphism of the network (Γ, r) if

$$r(\phi(x), \phi(y)) = r(x, y) \quad \text{for all } x \sim y.$$

Obviously, if $r(x, y) = 1$ for every $x \sim y$, then ϕ is an automorphism of Γ if and only if ϕ is an automorphism of (Γ, r). In the same way one defines an isomorphism between two different networks.

DEFINITION. Let ϕ be an automorphism of a graph Γ. We say that ϕ is bounded if

$$\sup_{x \in V(\Gamma)} d(x, \phi(x)) < \infty.$$

Given an automorphism ϕ of Γ, we say that a point $x \in V(\Gamma)$ is periodic if $\phi^n(x) = x$ for some integer $n \geq 1$, where ϕ^n is the n-th iterate of ϕ i.e., $\phi^0(x) = x$ and $\phi^n(x) = \phi(\phi^{n-1}(x))$ for $n > 0$. The results of this section are proved in Soardi and Woess (1991).

Theorem (4.8). *Let (Γ, r) be a locally finite, infinite network such that there exists a bounded automorphism ϕ of (Γ, r) with no periodic points. If there is a constant κ such that $r(x, y) \leq \kappa$ for every $x \sim y$, then $(\Gamma, r) \in \mathcal{O}_{\mathrm{HD}}$.*

Proof. Let u be a harmonic function on (Γ, r) such that

$$(4.9) \qquad u(z_1) - u(z_2) \to 0 \quad \text{as } z_1 \sim z_2 \text{ tend to infinity.}$$

We define a new function $f(x) = u(\phi(x)) - u(x)$ on (Γ, r). Since ϕ is a network automorphism we see that f is harmonic. Since Γ is connected and ϕ is bounded, there exists a number $K > 0$ independent of x such that $\phi(x)$ and x are joined by a path $\mathbf{p}(x)$ with less than K edges. Therefore $\phi(x) \to \infty$ as $x \to \infty$.

Now, f can be expressed as a sum, with bounded number of terms, of first differences of u along $\mathbf{p}(x)$. In other words, denoting by $x = x_0 \sim x_1 \sim \cdots \sim x_m =$

$\phi(x)$, $m \leq K$, the vertices of $\mathbf{p}(x)$, we have

$$|f(x)| \leq \sum_{j=1}^{m} |u(x_{j-1}) - u(x_j)|$$

(4.10)
$$\leq \sum_{j=1}^{m} r(x_{j-1}, x_j) \sum_{j=1}^{m} c(x_{j-1}, x_j)|u(x_{j-1}) - u(x_j)|^2$$

$$\leq K\kappa \sum_{j=1}^{m} c(x_{j-1}, x_j)|u(x_{j-1}) - u(x_j)|^2.$$

By (4.9) and (4.10), $f(x) \to 0$ as $x \to \infty$. By Corollary (1.37) we deduce that $f = 0$, so that $u(\phi(x)) = u(x)$ for all $x \in V$.

Let now x be fixed and let y be a neighbour of x. Then for every $n > 0$

(4.11) $u(x) - u(y) = u(\phi(x)) - u(\phi(y)) = \ldots = u(\phi^n(x)) - u(\phi^n(y)).$

Since ϕ has no periodic points $\phi^n(x) \to \infty$ and $\phi^n(y) \to \infty$ as $n \to \infty$. By (4.9) and (4.11) we obtain $u(x) = u(y)$ and, by connectedness, $u(x) = u(z)$ for all $z \in V$. Therefore u is constant. □

The problem now arises of investigating which networks fall within the scope of Theorem (4.8). Suppose that all the resistances are equal to 1. Then a first important example is provided by the graphs Γ which can be imbedded into $\mathbb{R}^n$ in such a way that the imbedding is periodic in some direction. By this we mean that there exists $v \in \mathbb{R}^n$ such that the map $x \mapsto x + v$ ($x \in V$) induces a graph automorphism. In particular, this condition is obviously satisfied by $\mathbb{Z}^n$ (the n–fold cartesian product of $\mathbb{Z}$, the two–ended infinite path represented by the integers; see Chapter 1).

More generally, every Cartesian product $\mathbb{Z} \times \Gamma$ has the bounded automorphism with no periodic points $\phi(n, x) = (n + 1, x)$, for all $n \in \mathbb{Z}$ and $x \in V$.

We now show that another important example is provided by the class of all vertex transitive graphs with polynomial growth whose combinatorial structure has been studied in detail by Trofimov (1985)(a). See Imrich and Seifter (1991) and Seifter (1991).

DEFINITION. We will say that Γ has polynomial growth if the number of vertices of Γ at a distance not more than n from some fixed vertex is bounded above by a polynomial in n.

The class of vertex transitive graphs with polynomial growth includes all lattices (in the sense of Trofimov (1984)) and the Cayley graphs of infinite, finitely generated nilpotent–by–finite groups.

Before stating the next result, let us recall the notion of the quotient graph with respect to an imprimitivity system.

An imprimitivity system of a vertex transitive group $G \leq \mathrm{Aut}(\Gamma)$ is a partition σ of the vertex set V of Γ into subsets called blocks, such that every element of G is a permutation of the blocks of σ. The quotient graph Γ^σ has vertex and edge sets given respectively by

$$V^\sigma = V/\sigma \qquad \text{(the blocks of } \sigma\text{)}$$
$$E^\sigma = \{[x^\sigma, y^\sigma] : x^\sigma \neq y^\sigma, [x,y] \in Y\},$$

where x^σ is the block containing x. There is a natural homomorphism from G into $\mathrm{Aut}(\Gamma^\sigma)$. The corresponding image of $\gamma \in G$ is denoted by γ^σ, and $G^\sigma = \{\gamma^\sigma : \gamma \in G\}$.

Theorem (4.12). *If Γ is vertex transitive and has polynomial growth, then there exists a bounded automorphism of Γ with no periodic points.*

Proof. By Theorem 2 in Trofimov (1985)(a) there exists an imprimitivity system σ of $\mathrm{Aut}(\Gamma)$ with finite blocks such that $H = \mathrm{Aut}(\Gamma^\sigma)$ is a finitely generated nilpotent–by–finite group and the stabilizer in H of a vertex in Γ^σ is finite. Moreover, $\gamma \in \mathrm{Aut}(\Gamma)$ is bounded on Γ if and only if γ^σ is bounded on Γ^σ.

For a given set A of generators of H, denote by $X_A = X(H, A)$ the Cayley graph of H with respect to A. Fix a vertex of Γ^σ and let J denote its stabilizer in H. By a result of Sabidussi (1964) there is a finite set of generators A of H such that $A \cap J = \emptyset$ and Γ^σ is isomorphic with X_A^τ, where τ denotes the imprimitivity system induced by the left cosets of J in $H = V(X_A)$.

We know by general group theory (see e.g. Kargapolov and Merzljakov (1979)) that there exists an infinite, torsion free, nilpotent subgroup $N \leq G^\sigma$ with finite index in H. In particular, N has nontrivial center.

Let ξ be an element in the center of N different from the identity e. Then $\{g^{-1}\xi g : g \in H\}$ is finite, and

$$\sup_{x^\sigma \in V(\Gamma^\sigma)} d(x^\sigma, \xi x^\sigma) \leq \sup_{g \in H} d(g, \xi g) = \sup_{g \in H} d(e, g^{-1}\xi g) < \infty$$

(in the first term d refers to the metric of Γ^σ, while it refers to the metric of X_A in the other two terms).

In other words, ξ is bounded on Γ^σ. Lifting ξ back to an automorphism ϕ of Γ, we have that ϕ is bounded.

Suppose now that $\phi^n(x) = x$ for some integer $n \geq 1$ and $x \in V(\Gamma)$. Then $\xi^\sigma = x^\sigma$ and, by the above, x^σ arises by isomorphism as a coset gJ, for some

$g \in H$. Therefore $\xi^n gJ = gJ$, so that $\xi^{nk} g = g$ for $k = |J|!$. Hence $\xi^{nk} = e$, a contradiction. This concludes the proof. □

Corollary (4.13). *If (Γ, r) is a network such that Γ is vertex transitive and has polynomial growth and r is identically 1, then $(\Gamma, r) \in \mathcal{O}_{\mathbf{HD}}$.*

We will see below (Theorem (4.75)) that uniqueness holds for a much larger class of vertex transitive graphs.

§3. Cartesian products

In this section we turn our attention to Cartesian products of graphs. Suppose that $\Gamma_1 = (V_1, Y_1)$ and $\Gamma_2 = (V_2, Y_2)$ are (infinite, connected) graphs. The Cartesian product $\Gamma_1 \times \Gamma_2$ is defined as the graph $\Gamma = (V, Y)$ such that

(4.14) $\qquad V = V_1 \times V_2$

$$Y = \{[(x_1, x_2), (y_1, y_2)] : \text{ either } x_2 = y_2 \text{ and } [x_1, y_1] \in Y_1$$
$$\text{or } x_1 = y_1 \text{ and } [x_2, y_2] \in Y_2\}.$$

The Cartesian product of n graphs is now defined recursively.

It turns out that uniqueness holds in cartesian products of locally finite networks with all the resistances equal to 1. This was proved by Thomassen (1989) who generalized a result by Soardi and Woess (1991).

We start with a lemma, also due to Thomassen, which is of independent interest. Recall that the notion of a canonical decomposition of a graph Γ was defined in §6 of Chapter III.

Lemma (4.15). *Let (Γ, r) be a locally finite network with all the resistances $r(x,y)$ equal to 1. Suppose that there exists a number κ and a canonical decomposition $\{V_n\}$ of Γ such that, for any two vertices $x, y \in V_n$, both joined to V_{n-1}, there is a path of length not larger than κn joining x and y in the subgraph induced by V_n. Then $(\Gamma, r) \in \mathcal{O}_{\mathbf{HD}}$.*

Proof. Suppose that $u \in \mathbf{HD}$ is not constant. There are $x_0 \sim x_1$ such that $\alpha = u(x_1) - u(x_0) > 0$. By harmonicity, we can define recursively an infinite path $\mathbf{p}_1$ with vertices $x_0 \sim x_1 \sim \cdots x_j \sim \cdots$ such that $u(x_j) < u(x_{j+1})$ for all $j \geq 0$. In the same way we can define an infinite path $\mathbf{p}_2$ with vertices $x_0 \sim x_{-1} \sim \cdots x_{-j} \sim \cdots$ such that $u(x_{-j}) < u(x_{-j+1})$ for all $j \geq 0$.

Let n be such that x_0 and x_1 belong to $V_{n-1} \cup V_n$. Every set V_m, with $m \geq n$ has nonvoid intersection with both paths $\mathbf{p}_1$ and $\mathbf{p}_2$. Therefore, for every $m \geq n$ there exist a vertex $a_{1,m}$ of $\mathbf{p}_1$ and a vertex $a_{2,m}$ of $\mathbf{p}_2$ joined by a path $\bar{\mathbf{p}}_m$ of length at most κm, all of whose vertices belong to V_m. Let $y_{m,0} \sim \cdots\cdots y_{m,s}$, with $s = s(m) \leq \kappa m$, be the vertices of $\bar{\mathbf{p}}_m$. Then

(4.16)
$$\alpha^2 < (u(a_{1,m}) - u(a_{2,m}))^2 = \left(\sum_{j=1}^{s} u(y_{m,j}) - u(y_{m,j-1})\right)^2$$
$$\leq \kappa m \sum_{j=1}^{s} |u(y_{m,j}) - u(y_{m,j-1})|^2.$$

Since the vertex sets of the paths $\bar{\mathbf{p}}_m$ are pairwise disjoint, we conclude from (4.16)

$$D(u) \geq \sum_{m=n}^{\infty} \sum_{j=1}^{s} |u(y_{m,j}) - u(y_{m,j-1})|^2$$
$$\geq \frac{\alpha^2}{\kappa} \sum_{m=n}^{\infty} \frac{1}{m}$$
$$= \infty$$

contradicting the assumption that u has finite Dirichlet sum. $\square$

Theorem (4.17). *Let G and H be infinite locally finite graphs. Let $\Gamma = G \times H$. Then the network (Γ, r), where r is identically 1, belongs to $\mathcal{O}_{\mathbf{HD}}$.*

Proof. We will show that there exists a canonical decomposition of Γ which satisfies the condition of Lemma (4.15). Choose a reference vertex (x_0, y_0) of Γ, where x_0 is a vertex of G and y_0 is a vertex of H. We define V_n in the following way: $(x, y) \in V_n$ if and only if

(4.18)
$$\max(d_G(x, x_0), d_H(y, y_0)) = n,$$

where d_G and d_H denote the distances in G and H respectively.

Let the distinct vertices (x, y) and (x', y') be both in V_n. Suppose, for instance, that $d_G(x, x_0) = n$. We have two possibilities: $d_G(x', x_0) = n$ or $d_H(y', y_0) = n$. Assume $d_G(x', x_0) = n$. Pick any vertex y^* in H such that $d_H(y_0, y^*) = n$. Remembering the definition (4.18) of V_n, we see that there is a path $\mathbf{p}_1$ with vertices in V_n, and length at most $2n$, joining (x, y) with (x, y^*). Similarly, there is a path $\mathbf{p}_2$ with vertices in V_n, and length at most $2n$, joining (x', y') with (x', y^*). Finally, there is a path $\mathbf{p}_3$, all of whose vertices are in V_n and whose length does not exceed $2n$, joining (x, y^*) with (x', y^*). It follows that there is a path $\mathbf{p}$ with vertices in V_n (contained in $\cup_{j=1}^{3} \mathbf{p}_j$) and length at most $6n$, which joins (x, y) with (x', y').

Suppose now that $d_H(y', y_0) = n$. Choose any x^* in G such that $d_G(x_0, x^*) = n$. Then there exists a path of length at most $2n$ and vertices in V_n joining (x^*, y')

with (x', y'). By the preceding argument, there exists a path of length not greater than $8n$, whose vertices are in V_n, which joins (x, y) with (x', y'). $\square$

REMARK. Paschke (1993)(b) defines a von Neumann dimension for discrete finitely generated groups which act properly on connected graphs. Such a dimension turns out to be zero if and only if the graph is in $\mathcal{O}_{HD}$. For instance, the von Neumann dimension is zero in the case of Cartesian products of infinite groups. See also Paschke (1993)(a).

§4. Nonuniqueness and ends

Now we will study sufficient conditions in order that a network does not belong to $\mathcal{O}_{HD}$. Such conditions will be formulated in terms of ends of a graph.

Let us recall briefly the notion of end of a graph. Let $\Gamma = (V, Y)$ be an infinite graph. For every finite subset L denote by $\Gamma(L)$ the subgraph of Γ induced by all vertices in V which are not in L. Let $\Gamma_j(L)$, $j = 1, 2, \ldots, \nu(L)$, $1 \le \nu(L) \le \infty$, be the infinite connected components of $\Gamma(L)$ (note that if Γ is locally finite, then $\nu(L) < \infty$). We say that two one–ended paths $\mathbf{p}_1$ and $\mathbf{p}_2$ are equivalent if, for every finite subset L, only finitely many vertices and edges of the two paths do not belong to to the same connected component $\Gamma_j(L)$. A class of equivalence of paths is called an end of Γ. The number of ends of Γ is defined by

$$(4.19) \qquad \sup\{\nu(L) : L \text{ a finite subgraph of } \Gamma\}.$$

Ends of locally finite graphs were introduced by Freudenthal (1944) and further studied by Halin (1964). For non locally finite graphs see Cartwright, Soardi and Woess (1993). It is also possible to define a topology on the ends space in such a way as to obtain a compactification of the graph.

We are interested in transient ends.

DEFINITION. We say that a network (Γ, r) has at least two transient ends if, for some finite subset L, there are at least two infinite connected components $\Gamma_1(L)$ and $\Gamma_2(L)$ of $\Gamma(L)$ such that the networks $(\Gamma_1(L), r_1)$ and $(\Gamma_1(L), r_2)$ are transient. Here r_1 and r_2 are the restrictions of r to the edge sets of $\Gamma_1(L)$ and $\Gamma_2(L)$ respectively.

Theorem (4.20). If (Γ, r) has at least two transient ends, then $(\Gamma, r) \notin \mathcal{O}_{HD}$.

Proof. Let L, $\Gamma_1(L)$ and $\Gamma_2(L)$ be as in the above definition. Let x be any vertex in L. For $j = 1, 2$, there is a finite path $\mathbf{p}_j$ with vertices

$$(4.21) \qquad x = x_0^j \sim x_1^j \sim \ldots \sim x_{n_j}^j = a_j$$

where a_j belongs to $\Gamma_j(L)$ and all the other vertices $x_2^j, \ldots, x_{n_j}^j$ are in L.

By Corollary (3.84)

$$(4.22) \qquad \lambda(\mathbf{P}_{a_1}^1) < \infty \quad \text{and} \quad \lambda(\mathbf{P}_{a_2}^2) < \infty,$$

where $\mathbf{P}_{a_1}^1$ and $\mathbf{P}_{a_2}^2$ denote the set of all infinite one–sided paths in $\Gamma_1(L)$ and $\Gamma_2(L)$, starting from a_1 and a_2, respectively.

For $j = 1, 2$, let $\mathbf{P}_x^j$ denote the set of all infinite one–sided paths starting at x, whose first n_j vertices are the x_k^j in (4.21) ($k = 0, \ldots, n_j$) and the remaining vertices are in $\Gamma_j(L)$.

We have

$$(4.23) \qquad \lambda(\mathbf{P}_x^j) < \infty \quad j = 1, 2.$$

Otherwise, by (3) of Lemma (3.75), there would exist a 1–chain I with finite energy such that

$$(4.24) \qquad \sum_{Y(\mathbf{p})} r(x, y)|I(x, y)| = \infty \quad \text{for all } \mathbf{p} \in \mathbf{P}_x^j.$$

In this case (4.24) would hold also for every $\mathbf{p} \in \mathbf{P}_{a_j}^j$, contradicting (4.22).

Now define a function f on the vertices of Γ by setting $f(x) = 1$ if x is a vertex of $\Gamma_1(L)$, $f(x) = 0$ otherwise. We have

$$D(f) \leq \sum_{x \in L} c(x) < \infty$$

by (1.1) and since L is finite. Let $f = u + v$, with $u \in \mathbf{D}_0$ and $v \in \mathbf{HD}$, be the Royden decomposition (3.69) of f. By Theorem (3.86) $u(x)$ tends to 0 along the vertices of almost all paths in $\mathbf{P}_x^1$, so that $v(x) \to 1$ along the vertices of the same paths. Note that, by (4.23) such a set of paths is not void. For the same reason, $u(x) \to 0$ along the vertices of almost all paths in $\mathbf{P}_x^2$, so that $v(x) \to 0$ on the same paths. By (4.23) again such a set of paths is not void. Therefore v is nonconstant. $\square$

Roughly speaking, the above Theorem asserts that "joining" together two transient networks produces a network which does not belong to $\mathcal{O}_{\mathbf{HD}}$. Transient trees are typical examples of of this situation.

Corollary (4.25). *Let T be a locally finite transient tree such that for some $\delta > 0$*

$$r(x, y) \geq \delta \quad \text{for all } x \sim y.$$

Then T does not belong to $\mathcal{O}_{HD}$.

Proof. Fix any vertex x_0 and let R be the resistance between x_0 and ∞. The expression of R was computed in Chapter III, equation (3.58). Namely, $R = \lim_{n \to \infty} R_n$, where

$$R_n = \cfrac{1}{\sum\limits_{x_1 \in U(x_0)} r(x_0, x_1) + \cfrac{1}{\sum\limits_{x_2 \in U(x_1)} r(x_1, x_2) + \cfrac{1}{\ddots + \cfrac{1}{\sum\limits_{x_n \in U(x_{n-1})} r(x_{n-1}, x_n)}}}}$$

Recall that $U(x_n)$ is the set of all neighbors y of x_{n-1} such that $y \prec x_{n-1}$ (see Chapter II, §2).

For every vertex z let τ_z denote the infinite subtree with root z induced by all the vertices $x \in T$ such that $x \prec z$. The corollary will follow from Theorem (4.20) and from (3.58) if we prove that there are at least two distinct vertices z_1 and z_2 with the following properties: τ_{z_1} and τ_{z_2} are transient and both the relations $z_1 \prec z_2$ and $z_2 \prec z_1$ are false.

Assume, by way of contradiction, that we cannot find two such vertices. Set $y_0 = x_0$. There is a unique $y_1 \in U(y_0)$ such that τ_{y_1} is transient. We have $R = r(y_0, y_1) + \rho_1$, where ρ_1 is the resistance between y_1 and ∞ of τ_{y_1}. Analogously, there exists a unique $y_2 \in U(y_1)$ such that τ_{y_2} is transient; as before and $\rho_1 = r(y_0, y_2) + \rho_2$, where ρ_2 is the resistance between y_2 and ∞ of τ_{y_2}.

Carrying on this process, for every n we find a unique $y_n \in U(y_{n-1})$ such that τ_{y_n} is transient and $\rho_{n-1} = r(y_{n-1}, y_n) + \rho_n$, where ρ_n is a positive number with the analogous meaning as above. Since $r(x, y) \geq \delta$ we have for all n

$$R = \sum_{j=1}^{n} r(y_{n-1}, y_n) + \rho_n \geq n\delta.$$

Therefore $R = \infty$, which is absurd. $\square$

We will return to uniqueness and ends of general graphs in §7.

§5. The strong isoperimetric inequality

Let $\Gamma = (V, Y)$ be a graph. In the sequel L will always denote a finite nonempty subset of V. The following definition of boundary of L is implicitly contained in §5 of Chapter I.

DEFINITION. Suppose that L is a finite subset of vertices of Γ. We will denote by $\daleth(L)$ ("daleth") the (combinatorial) boundary of L, defined as the subset of all vertices in L which have a neighbour x not in L (i.e. the set of all (combinatorial) boundary points of L; see §5 of Chapter I).

DEFINITION (see e.g. Dodziuk (1984), Varopoulos (1985), Gerl (1988), Ancona (1988)). We say that a locally finite graph Γ satisfies a strong isoperimetric inequality if there is a constant $\kappa > 0$ such that

(4.26) $|\daleth(L)| \geq \kappa|L|$ for every finite subset L,

where $|\cdot|$ denotes cardinality.

It is immediately seen that if Γ satisfies a strong isoperimetric inequality, then Γ has exponential growth: denoting by B_n the ball of radius n centered at any vertex of Γ, we have that $|B_n| \geq (1 + \kappa)^n$.

We will show in Theorem (4.27) that a graph satisfying a strong isoperimetric inequality is transient in rather a strong way. However, there are recurrent graphs (hence not satisfying a strong isoperimetric inequality) with exponential growth: see Chapter V.

Strong isoperimetric inequalities are related to several important properties of random walks on graphs. Namely, let Γ be any locally finite graph with all the resistances equal to 1. We know from Lemma (3.12) that P is a bounded hermitian operator on ℓ^2 with norm less than or equal to 1. Furthermore, if the degrees of the vertices of Γ are uniformly bounded, then it is easily seen that there exists a constant k such that for every finitely supported function f we have $D(f) \leq k\|f\|_2^2$.

If the graph satisfies a strong isoperimetric inequality, we can say much more.

Theorem (4.27). *Let $\Gamma = (V, Y)$ be a graph such that $\sup_{x \in V} \deg(x) < \infty$. Then the following are equivalent*

(1) *Γ satisfies a strong isoperimetric inequality;*

(2) *there exists a constant γ such that for every finitely supported f*

(4.28) $\|f\|_2^2 = \sum_{x \in V} c(x)|f(x)|^2 \leq \gamma D(f)$ *where $c(x) = \deg(x)$;*

(3) *the norm of P as an operator on ℓ^2 is smaller than 1;*

(4) *there exists a number σ, $0 < \sigma < 1$, such that $p^n(x, y) = o(\sigma^n)$ for all $x, y \in V$;*

(5) *G is a bounded operator on ℓ^2.*

Proof. For the proof we follow Varopoulos (1985), Ancona (1988) and Gerl (1988).

$(1) \Rightarrow (2)$. Set, for every finitely supported function f,

$$\|f\|_1 = \sum_{x \in V} \deg(x)|f(x)|$$

$$\|f\|_S = \frac{1}{2} \sum_{x \sim y} |f(x) - f(y)|.$$

Then, inequality (4.26) implies

$$(4.29) \qquad M^{-1}\kappa\|e_L\|_1 \leq 2\|e_L\|_S,$$

where e_L is the characteristic function of the finite set L and $M = \max_{x \in V} \deg(x)$.

Let f be any finitely supported nonnegative function on V. Let $a_1 < \ldots < a_n$ be the values taken by f. Then f can be written as $f(x) = \sum_{j=1}^{n} a_j e_{L_j \setminus L_{j+1}}$, where L_j is the set of all $x \in V$ such that $f(x) \geq a_j$, and $L_{n+1} = \emptyset$. Then, rearranging terms,

$$f(x) = \sum_{j=1}^{n} c_j e_{L_j}, \quad c_1 = a_1, \text{ and } c_j = a_j - a_{j-1} \text{ for } j = 2, \ldots, n.$$

Since $(e_{L_j}(x) - e_{L_j}(y))(e_{L_i}(x) - e_{L_i}(y)) \geq 0$ for every i and j, we get immediately from (4.29)

$$\|f\|_1 \leq \sum_{j=1}^{n} c_j\|e_{L_j}\|_1 \leq \kappa_1 \sum_{j=1}^{n} c_j\|e_{L_j}\|_S$$

$$= \kappa_1\|\sum_{j=1}^{n} c_j e_{L_j}\|_S$$

$$= \kappa_1\|f\|_S$$

where we set $\kappa_1 = 2M\kappa^{-1}$. Therefore, for every finitely supported function f

$$(4.30) \qquad \|f\|_1 \leq \kappa_1\|\,|f|\,\|_S \leq \kappa_1\|f\|_S.$$

Applying now (4.30) to f^2 and using Schwartz inequality we have

$$\|f\|_2^2 \leq \kappa_1 \sum_{x \sim y} |f^2(x) - f^2(y)|$$

$$(4.31) \qquad \leq \kappa_1 \left(\sum_{x \sim y} |f(x) - f(y)|^2\right)^{1/2} \left(\sum_{x \sim y} |f(x) + f(y)|^2\right)^{1/2}.$$

We have

$$(4.32) \qquad \sum_{x \sim y} |f(x) + f(y)|^2 \leq 4\|f\|_2^2.$$

Thus (2) follows from (4.31) and (4.32).

(2) $\Rightarrow$ (1) . This is immediate by taking in (4.28) $f = e_L$.

(2) $\Rightarrow$ (3). Let $(\cdot, \cdot)$ denote the inner product in ℓ^2. For all real valued finitely supported functions f and g we set as in (3.65)

$$
\begin{aligned}
\lfloor f, g \rfloor &= (f, (1 - P)(g)) \\
&= \frac{1}{2} \sum_{x \sim y} (f(x) - f(y))(g(x) - g(y)).
\end{aligned}
$$

Therefore

$$D(f) = \|f\|_2^2 - (f, P(f)) \quad \text{for all finitely supported } f.$$

Since P is self–adjoint as an operator on ℓ^2, we have, denoting by α the norm of P,

(4.33) $$\alpha = \sup(f, P(f)),$$

where the supremum in (4.33) is taken over all the real valued finitely supported functions f with ℓ^2 norm equal to 1. Thus we get

$$
\begin{aligned}
\alpha &= \sup \frac{(f, P(f))}{\|f\|_2^2} = 1 - \inf \frac{D(f)}{\|f\|_2^2} \\
&\leq 1 - \gamma^{-2}
\end{aligned}
$$

(3) $\Rightarrow$ (4). Set $s = \limsup_{n \to \infty} (p^n(x, y))^{1/n}$, where P^n have the same meaning as in §4 of Chap I. We have

$$p^n(x, y) \leq (P^n(\delta_x), \delta_y) \leq \|P^n\|,$$

where $\|P^n\|$ is the operator norm of P^n. Then $s \leq \limsup_{n \to \infty} \|P^n\|^{1/n} = \alpha$ (it is also clear that s does not depend on x and y). Then (4) follows with any σ such that $1 > \sigma > \alpha$.

(4) $\Rightarrow$ (3). In the following lines f is finitely supported with ℓ^2 norm equal to 1. The supremum is taken with respect to f and we use $\|\cdot\|$ again to denote the operator norm.

We have

$$\|P\|^2 = \sup(P(f), P(f)) = \sup(P^2(f), f) \leq \|P^2\|$$

so that

(4.34) $$\|P^2\| = \|P\|^2.$$

Set $S_n = (P^n(f), P^n(f)) = (P^{2n}(f), f)$, for $n = 0, 1, \ldots$. For all real x

$$0 \leq (P^{n-1}(f) + xP^{n+1}(f), P^{n-1}(f) + xP^{n+1}(f))$$
$$= S_{n-1} + 2xS_n + x^2 S_{n+1}.$$

It follows $S_n^2 \leq S_{n-1}S_{n+1}$, whence

$$S_1 = \frac{S_1}{S_0} \leq \frac{S_2}{S_1} \leq \cdots \frac{S_n}{S_{n-1}} \leq \cdots$$

Thus we have $S_1 \leq S_n^{1/n}$. Then

(4.35) $(P^2(f), f) \leq (P^{2n}(f), (f))^{1/n}.$

Let U be the support of f and let c be the cardinality of U. Then

(4.36)
$$(P^{2n}(f), (f)) = \sum_{x,y} \deg(x) p^{2n}(x, y) f(x) f(y)$$
$$\leq c^2 \max_x |f(x)|^2 \max_x \deg(x) \max_{x,y \in U} p^{2n}(x, y).$$

By (4.34), (4.35) and (4.36) we obtain

$$\|P\|^2 = \sup_{\|f\|_2=1} (P^2(f), f) \leq \sigma^2 < 1.$$

$(3) \Rightarrow (5)$. Let $\|P\| = \alpha < 1$. Then the series $\sum_{n=0}^{\infty} P^n$ converges absolutely (in the operator norm) to an operator T which is the inverse $(1 - P)^{-1}$ of $(1 - P)$. Therefore G is a bounded operator on ℓ^2.

$(5) \Rightarrow (2)$. As above we have for every real valued finitely supported f and g

$$\lfloor f, g \rfloor = (f, (1 - P)(g)) = \frac{1}{2} \sum_{x \sim y} (f(x) - f(y))(g(x) - g(y)).$$

Then (the functions g in the following lines have ℓ^2 norm equal to 1 and the supremum is taken with respect to g)

$$\|f\|_2 = \sup(f, g) = \sup(f, (1 - P)G(g)) = \sup \lfloor f, G(g) \rfloor.$$

Therefore
$$\|f\|_2 \leq \sup \lfloor f, f \rfloor^{1/2} \lfloor G(g), G(g) \rfloor^{1/2}$$
$$= (f, (1 - P)(f))^{1/2} (G(g), g)^{1/2}.$$

Now $(f, (1 - P)(f)) = D(f)$, and $(G(g), g) \leq K$ where K is the norm of G. Thus (2) holds. $\square$

REMARK. Inequality (4.28) is known as the Sobolev–Dirichlet inequality. An obvious consequence of (4) is that the associated simple random walk on Γ is transient. In fact, a random walk satisfying (4) is called strongly transient.

A necessary and sufficient condition for transience is given by the so called Beurling inequality (see Beurling and Deny (1959)): for all $x \in V$ there exists a constant $c = c(x)$ such that

$$|f(x)| \leq c(x)D(f) \quad \text{for all } f \in \ell_0;$$

see Varopoulos (1985) and Kalpazidou (1991).

The notion of strong isoperimetric inequality and the results of Theorem (4.27) have been extended to general networks (Markov chains) by Kaimanovic (1992).

By Lemma (3.3) we know that, as u varies in $\mathbf{D}$, the range of $(1 - P)(u)$ is contained in ℓ^2. We can say more if Γ satisfies a strong isoperimetric inequality.

Corollary (4.37). *Let Γ be a transient locally finite graph with all the resistances equal to 1. Then $(1 - P)$ maps $\mathbf{D}$ onto ℓ^2 if and only if Γ satisfies a strong isoperimetric inequality.*

Proof. If Γ satisfies a strong isoperimetric inequality, then, by (4.28), $D(f)^{1/2}$ and $\|f\|_2$ are equivalent norms on the dense subspace of all finitely supported functions. Hence $\mathbf{D}_0 = \ell^2$. Then, for every $g \in \ell^2$, $(1 - P)^{-1}(g) \in \mathbf{D}_0$.

To prove the converse observe that, by the Royden Decomposition Theorem, for every $f \in \mathbf{D}$ there exists $v \in \mathbf{D}_0$ such that $(1 - P)(f) = (1 - P)(v)$. Furthermore, $(1 - P)$ is one–to–one on $\mathbf{D}_0$. If $(1 - P)$ maps $\mathbf{D}_0$ onto ℓ^2, then, by the open mapping theorem, the norms in ℓ^2 and in $\mathbf{D}_0$ are equivalent. Hence (4.28) holds and Γ satisfies a strong isoperimetric inequality. $\square$

Examples of graphs which satisfy a strong isoperimetric inequality include edge graphs of certain tilings of the plane, in particular triangulations such that every vertex has degree at least 7 (Dodziuk (1984), Mohar (1988), Soardi (1990)) They include also trees such that there is a finite upper bound for the length of all unbranched paths, in particular homogeneous trees of degree greater than 2 (Gerl (1986); see Theorem (5.54) in the next chapter. Furthermore, the Cayley graphs of many Fuchsian groups acting the hyperbolic upper half–plane or the hyperbolic disk satisfy a strong isoperimetric inequality; see §6 below.

It was proved in Soardi and Woess (1990) that a locally finite vertex transitive graph with infinitely many ends satisfies a strong isoperimetric inequality. Therefore we may state the following Corollary of Theorem (4.20).

Corollary (4.38). *Let (Γ, r) be a locally finite network with all the resistances 1. Assume that Γ is vertex transitive and has infinitely many ends. Then $(\Gamma, r) \notin \mathcal{O}_{HD}$.*

Proof. There exists a finite connected subset L of Γ such that $\Gamma(L)$ splits into $\nu \geq 2$ infinite connected components $\Gamma_1(L), \ldots, \Gamma_\nu(L)$. Since Γ satisfies a strong isoperimetric inequality, every component $\Gamma_j(L)$ has the same property. Hence the $\Gamma_j(L)$ are transient. The Corollary now follows from Theorem (4.20). □

REMARK. A vertex transitive graph has one, two or infinitely many ends (see Halin (1973)). It was proved by Imrich and Seifter (1988/89) that a vertex transitive graph with exactly two ends has linear growth (a ball of radius n centered at x_0 contains at most const.n edges). Hence (Γ, r) with r identically 1, is recurrent by Nash–Williams Theorem. Such networks are in $\mathcal{O}_{HD}$ by Theorem (4.6) or Theorem (4.12).

It is proved in Bekka and Valette (1994) that, if Γ is the Cayley graph of a finitely generated group, and if Γ satisfies the strong isoperimetric inequality, then $HD/\mathbb{C}$ is isomorphic to the first cohomology group with coefficients in the left regular representation. Several result concerning uniqueness are then deduced. For instance, $\Gamma \in \mathcal{O}_{HD}$ if the group has Kazhdan's property (T), or, more generally, if the first L^2 Betti number is 0.

In the next section we will show that there exist networks $(\Gamma, r) \notin \mathcal{O}_{HD}$ with $r = 1$ and Γ vertex transitive and with only one end. Therefore if Γ is vertex transitive and has only one end, then the network may (see e.g. Theorem (4.17)) or may not belong to $\mathcal{O}_{HD}$. Finally, we notice that the structure of vertex transitive graphs with infinitely many ends was studied by Nevo (1991).

The following definition is essentially due to Thomassen (1989), (1990). See also Doyle (1988).

DEFINITION. A locally finite graph $\Gamma = (V, Y)$ has moderate growth if there exists an sequence of finite sets U_n such that $U_n \subseteq U_{n+1}$, $\cup_{n=0}^{\infty} U_n = V$, and

$$\lim_{n \to \infty} \frac{|\daleth(U_n)|}{|U_n|} = 0.$$

REMARK. Examples of graphs of moderate growth include all the graphs with polynomial growth, and more generally, all the graphs with subexponential growth, i.e. graphs such that, for every $\kappa > 0$, $\liminf_{n \to \infty} |B_n|(1 + \kappa)^{-n} = 0$ (where B_n the ball of radius n centered at any fixed vertex of Γ). This can be seen as follows. Set $U_0 = \{o\}$ where o is any fixed vertex. Suppose that we have defined $U_j = B_{n_j}$ for $0 < j < k$ in such a way that $|\daleth(U_j)|/|U_j| < 1/j$. Since the graph grows subexponentially, there exists a ball B_{n_k} of radius $n_k > n_{k-1}$ such that $|\daleth(B_{n_k})|/|B_{n_k}| < 1/k$ (otherwise $|B_n| \geq \text{constant}(1 + 1/k)^n$ for n large enough). So, it suffices to set $U_k = B_{n_k}$. It is also clear that for every k we have $U_{k-1} \subset U_k$ and also $\cup_{k=0}^{\infty} U_k = V$.

DEFINITION. Let $\Gamma = (V, Y)$ be an infinite graph and let $\text{Aut}(\Gamma)$ be the automorphism group of Γ. An orbit of $\text{Aut}(\Gamma)$ on V is a subset K of V such that

i) for every $\phi \in \text{Aut}(\Gamma)$, $\phi(K) = K$.

ii) for every $x, y \in K$ there exists an automorphism ϕ on Γ such that $\phi(x) = y$.

If a graph has moderate growth, then it does not satisfy a strong isoperimetric inequality. The converse implication was proved by Thomassen (1989) for vertex transistive graphs and extended to the case where there are only finitely many orbits by Medolla and Soardi (1993).

Theorem (4.39). *Let (Γ, r) be an infinite graph and assume that $\text{Aut}(\Gamma)$ acts on V with a finite number of orbits. Then Γ has moderate growth if and only if it does not satisfy a strong isoperimetric inequality.*

Proof. Suppose that Γ does not satisfy (4.26). Denote by $K_1, K_2, \ldots, K_s$ the orbits of $\text{Aut}(\Gamma)$ on V. Fix a vertex v_i in each orbit K_i. We set $U_0^{(1)} = \{v_1\}, U_0^{(2)} = \{v_2\}, \ldots, U_0^{(s)} = \{v_s\}$. Let $d_i = \deg(x)$ for all $x \in K_i$ and put $d = \max\{d_1, \ldots, d_s\}$.

Suppose now that we have already defined $U_n^{(i)}$ for every $i = 1, \ldots, s$ and let $m_n^{(i)}$ be the maximum distance from v_i to a vertex in $U_n^{(i)}$. Set

$$m_n = \max\{m_n^{(1)}, m_n^{(2)}, \ldots, m_n^{(s)}\}$$

and choose any positive number $\alpha < d^{-(m_n+1)}$. By assumption there exists a finite set $L \subset V$ such that $|\daleth L| < \alpha|L|$. Moreover, there exists a vertex $o \in L$ whose distance from the boundary of L is strictly larger than m_n. Otherwise we would have

$$|L| \leq |B(o, m_n)| \leq d^{m_n+1}|\daleth L| \leq \alpha^{-1}|\daleth L|$$

which is absurd. Since the vertex o belongs to K_j for some j, we may suppose $o = v_j$. We set $U_{n+1}^{(j)} = L$, and $U_{n+1}^{(h)} = U_n^{(h)}$ if $h \neq j$. Since $d(o, \daleth L) > m_n$ we have $U_{n+1}^{(j)} \supset B(o, m_n) \supseteq B(o, m_n^{(j)}) \supseteq U_n^{(j)}$. Note that m_n increases at each step, since $m_{n+1} \geq m_{n+1}^{(j)} \geq d(o, \daleth U_{n+1}^{(j)}) = d(o, \daleth L) > m_n$.

Carrying on this process, we construct s nondecreasing sequences of finite subsets of vertices. Among the sequences $\{U_n^{(1)}\}, \ldots, \{U_n^{(s)}\}$ defined in such a way, there must exist at least one that contains infinitely many distinct terms. For instance, let $\{U_n^{(1)}\}$ be such a sequence. Passing to a subsequence if necessary, we may assume that the sequence $\{U_n^{(1)}\}$ does not contain repeated terms.

Since $U_n^{(1)}$ contains a ball centered at v_1 whose radius m_{n-1} increases at each step, $\bigcup_{n=0}^{\infty} U_n^{(1)} = V$. Finally, by construction,

$$\frac{|\daleth U_n^{(1)}|}{|U_n^{(1)}|} < \frac{1}{d^{m_n+1}}$$

and the term on the left converges to zero as n tends to infinity. Thus the sequence $\{U_n^{(1)}\}$ satisfies all the requirements of the definition of moderate growth. $\square$

§6. Graphs embedded in the hyperbolic disk

In this section we prepare ourselves for a result of Cartwright and Woess (1992). This result states that if Γ is uniformly embedded in the hyperbolic disk Ω and satisfies a strong isoperimetric inequality, then (Γ, r), with r identically 1, does not belong to $\mathcal{O}_{HD}$. The class of graphs satisfying these assumptions contains many vertex–transitive graphs with only one end, as we shall see.

DEFINITION. A graph $\Gamma = (V, Y)$ is said to be uniformly embedded in Ω if $V \subset \Omega$ and there is a constant c such that

(4.40) $c^{-1}\rho(z_1, z_2) \leq d(z_1, z_2) \leq c\rho(z_1, z_2)$ for all $z_1, z_2 \in V$

where ρ denotes the hyperbolic distance between z_1 and z_2.

In order to make more precise the scope of Cartwright and Woess' result we will prove in an elementary way that the edge graph of the tesselation of the hyperbolic upper half–plane $A = \{z \in \mathbb{C} : \Im z > 0\}$, associated with the fundamental region of a cocompact Fuchsian group, satisfies a strong isoperimetric inequality and is uniformly embedded in A (i.e., the analogue of (4.40) holds). We refer e.g. to the book of Magnus (1974) for examples and pictures of edge graphs of tesselations of A or Ω.

Let ρ denote the distance on the hyperbolic upper half–plane as well. Then

(4.41) $$\tanh\left(\frac{1}{2}\rho(z, w)\right) = \frac{|z - w|}{|z - \overline{w}|}$$

(see e.g. Beardon (1983) for general facts on the hyperbolic disk and the hyperbolic upper half–plane).

It follows easily that a hyperbolic circle of radius r centered at $x + iy$ is in fact an euclidean circle of radius $2y\alpha/(1 - \alpha^2)$ centered at $x + iy(1 + \alpha^2)/(1 - \alpha^2)$, where we made $\alpha = \tanh(\frac{1}{2}r)$.

Writing $z = x + iy$, the hyperbolic area $\text{meas}(C)$ of a region C of the upper half–plane is given by the integral

$$(4.42) \qquad \text{meas}(C) = \int_C \frac{dxdy}{y^2}.$$

In particular, the measure of a hyperbolic circle depends only on the radius r but not on the centre. Such a measure will be denoted by $\sigma(r)$. See equation (4.53) below for the expression of $\sigma(r)$ in the hyperbolic disk.

Assume that G is a cocompact Fuchsian group acting on A. Then there exists a compact convex hyperbolic polygon F which is a fundamental region for G; furthermore the family $\{gF\}_{g \in G}$ is a locally finite tiling of A (see Beardon (1983), chap 9). See §7 of chap V for precise definitions concerning tilings.

Theorem (4.43). *Suppose that Γ is the edge graph of the tiling $\{gF\}_{g \in G}$, where G is a cocompact Fuchsian group acting on the hyperbolic upper half–plane. Then Γ is uniformly embedded in the hyperbolic upper half–plane.*

Proof. Since F is compact there are hyperbolic circles B_1 and B_2 of radius r_1 and r_2 such that $B_1 \subseteq F \subseteq B_2$. Therefore $gB_1 \subseteq gF \subseteq gB_2$ for all $g \in G$.

Let now z and w be distinct vertices of Γ. Then

$$(4.44) \qquad \rho(z, w) \geq M \quad \text{for some } M > 0.$$

Let γ be the geodesic arc joining z and w. Let n be the number of tiles intersecting γ. Denoting by h the number of edges of F we have

$$(4.45) \qquad d(z, w) \leq hn.$$

Denote by C the union of all hyperbolic circles of radius $2r_2$ centered at the points of γ. Then

$$(4.46) \qquad n \leq \frac{\text{meas}(C)}{\sigma(r_1)}.$$

We have to compute $\text{meas}(C)$. We can map isometrically γ onto a line segment of the imaginary axis by means of a suitable Möbius transformation g. Let $ia = g(z)$ and $ib = g(w)$, with $a < b$. Set $\alpha = \tanh(r_2)$. We have from (4.42)

$$(4.47) \qquad \begin{aligned} \text{meas}(g(C)) &\leq 2 \int_{a(1-\alpha)/(1+\alpha)}^{b(1+\alpha)/(1-\alpha)} \frac{dy}{y} \int_0^{2\alpha/(1-\alpha^2)} dx \\ &= \frac{4\alpha}{(1-\alpha^2)}\big((\log b - \log a) + \kappa\big), \end{aligned}$$

where $\kappa = 4\log(1 + \alpha)/(1 - \alpha)$.

The transformation g preserves hyperbolic length and areas, so that $\mathrm{meas}(C) = \mathrm{meas}(gC)$ and $\rho(z, w) = \log b - \log a$. Therefore, by (4.44) and (4.47) we have $\mathrm{meas}(C) \leq \mathrm{const}.\rho(z, w)$. Thus we get $d(x, w) \leq c\rho(z, w)$ from (4.45) and (4.46), with $c = \mathrm{const}.h/\sigma(r_1)$.

To complete the proof we have only to observe that, by the definition of r_2, $2r_2 d(z, w) \geq \rho(z, w)$ for every couple of vertices z and w of Γ. $\square$

Lemma (4.48). *Let Γ be as in Theorem (4.43). Let $z_1, \ldots, z_n$ be the vertices of F and let d_j, $j = 1, \ldots, n$, be the degree of z_j. Then*

$$2\sum_{j=1}^{n} d_j^{-1} < n - 2.$$

Proof. The proof is based on Siegel (1971), pages 44–45. Let us divide the vertices of F into equivalence classes, two vertices w and z being equivalent if $z = g(w)$ for some $g \in G$. Let z be a vertex of F and let d be the degree of z. Denote by $g_1 F, \ldots, g_d F$ the regions which meet at z. The stabilizer G_z of z permutes the regions $g_j F$. Hence G_z is a finite cyclic subgroup of G of order, say, m. We set $G_z = \{h_1, \ldots, h_m\}$.

Setting $w_j = g_j^{-1}(z)$, we have that w_j is a vertex of F. Such vertices need not being distinct; in fact we are about to show that they are equal in groups of m. Specifically, for a fixed k, the m different mappings $h_1 g_k, \ldots, h_m g_k$ carry w_k into z, and the m corresponding distinct regions $F_{h_1 g_k}, \ldots, F_{h_m g_k}$ all have z in common. It follows that m of the points w_j coincide with w_k. Conversely, the equality $w_j = w_k$ implies that $g_k^{-1} g_j$ belongs to G_z.

Let ω_j be the angle subtended by z in $g_j F$. The same argument as above shows that the angles ω_j are equal in groups of m. Since $\omega_1 + \cdots \omega_d = 2\pi$, we have that the sum of the angles subtended by the vertices equivalent to z in F is $2\pi/m$.

Let $\mathcal{V}$ be a maximal set of pairwise inequivalent vertices of F. For all $v \in \mathcal{V}$ let m_v be the order of the stabilizer G_v, and let d_v be the degree of v. Finally, let $s_v = d_v/m_v$ be the cardinality of the equivalence class of v. We then have

$$\sum_{j=1}^{n} 2\pi d_j^{-1} = \sum_{v \in Cal\mathcal{V}} 2\pi s_v d_v^{-1}$$

$$= \sum_{v \in \mathcal{V}} 2\pi/m_v$$

$$< \pi(n - 2)$$

since $\pi(n - 2) - \sum_{v \in \mathcal{V}} 2\pi/m_v$ is the hyperbolic area of F. $\square$

Theorem (4.49). *Let Γ be as in Theorem (4.43). Then Γ satisfies a strong isoperimetric inequality.*

Proof. The proof runs along the lines of Dodziuk's proof (1984) that the edge graph of a triangulation of the plane satisfies a strong isoperimetric inequality, provided that all the vertices have degree not smaller than 7 (see Dodziuk (1984)).

Since in the proof we must introduce several finite subgraphs of Γ, say Υ, we agree to denote by $V(\Upsilon)$ and $E(\Upsilon)$ the respective vertex and edge sets. If an edge is in $E(\Upsilon)$, then both the endpoints are in $V(\Upsilon)$, while isolated vertices are possible in $V(\Upsilon)$.

Let Γ' be a finite subgraph of Γ induced by its vertex set. We will write Γ' as the union of three finite subgraphs. First let $T(\Lambda)$ denote the set of tiles all of whose edges are in $E(\Gamma')$. Let Λ denote the union of all the edges of the tiles in $T(\Lambda)$. Let Λ_0 denote the set of all isolated points of Γ' and let Λ_1 be the union of all the edges of Γ' which are not edges of a tile in $T(\Lambda)$. Then $\Lambda, \Lambda_0, \Lambda_1$ are finite subgraphs of Γ'. We have $\Gamma' = \Lambda \cup \Lambda_0 \cup \Lambda_1$.

Finally, let $\partial\Lambda$ denote the union of all the edges of Λ which are edges of only one tile in $T(\Lambda)$. Note that Λ is not necessarily connected. In any case we may apply Euler's formula (see e.g. Bollobàs (1979) so as to have

$$-|E(\Lambda)| + |T(\Lambda)| + |V(\Lambda)| = 2 - q,$$

where q is the number of connected components of Λ. Then

(4.50) $|E(\partial\Lambda)| \geq q - 2 = |E(\Lambda)| - |T(\Lambda)| - |V(\Lambda)|.$

Let, as in the proof of Lemma (4.48), $\mathcal{V}$ be a maximal set of pairwise inequivalent vertices of F. For every $z \in V$ let d_z denote the degree of z and let s_z be the number of vertices of F equivalent to z. For every $z \in \mathcal{V}$, let $V_z(\Lambda)$ and $V_z(\partial\Lambda)$ denote the vertices in Λ and $\partial\Lambda$, respectively, which are of the form $g(z)$ with $g \in G$. A simple counting argument shows that

$$s_z|T(\Lambda)| \geq d_z \left(|V_z(\Lambda)| - |V_z(\partial\Lambda)|\right) + |V_z(\partial\Lambda)| \quad \text{for every } z \in \mathcal{V}.$$

Now, $|V(\partial\Lambda)| \leq 2|E(\partial\Lambda)|$ and $|V_z(\partial\Lambda)| \leq |V(\partial\Lambda)|$. Summing over z we get

(4.51) $$|V(\Lambda)| \leq |T(\Lambda)| \sum_{z \in \mathcal{V}} \frac{s_z}{d_z} + 2|E(\partial\Lambda)| \sum_{z \in \mathcal{V}} (1 - d_z^{-1}).$$

Since every edge belongs to at most two tiles in $T(\Lambda)$, we also have $n|T(\Lambda)| \leq 2|E(\Lambda)|$, where n is the number of edges of F. It follows from inequalities (4.50) and (4.51)

(4.52) $$|E(\partial\Lambda))|\left(1 + 2\sum_{z \in \mathcal{V}}(1 - d_z^{-1})\right) \geq |T(\Lambda)|\left(\frac{n-2}{2} - \sum_{z \in \mathcal{V}} \frac{s_z}{d_z}\right).$$

If $z_1, \ldots, z_n$ is an enumeration of the vertices of F, then

$$\sum_{z \in \mathcal{V}} \frac{s_z}{d_z} = \sum_{j=1}^{n} d_j^{-1},$$

so that the left–hand side of (4.52) is positive by Lemma (4.48). As $|V(\Lambda)| \leq n|\mathcal{T}(\Lambda)|$ and $2|E(\partial\Lambda)| \leq m|V(\partial\Lambda)|$, where m is the maximum degree d_z, we get

$$|V(\Lambda)| \leq mn \frac{1 + 2 \sum_{z \in \mathcal{V}} (1 - d_z^{-1})}{n - 2 - 2 \sum_{j=1}^{n} d_j^{-1}} |\daleth(V(\Lambda))|.$$

Now, we evaluate $\daleth(V(\Lambda_1))$. If x is an interior vertex of $V(\Lambda_1)$, then there must be a boundary vertex whithin distance $[n]/2$ from x. Therefore

$$|\daleth(V(\Lambda_1))| \geq ([n]/2)^{-m} |V(\Lambda_1)|.$$

It follows that

$$|\daleth(V(\Gamma'))| \geq \frac{1}{2}([n]/2)^{-m}(|V(\Lambda_0)| + |V(\Lambda_1)| + |V(\partial\Lambda)|).$$

As $|V(\Gamma')| \leq |V(\Lambda_0)| + |V(\Lambda_1)| + |V(\Lambda)|$, we get $|\daleth(V(\Gamma'))| \geq \kappa |V(\Gamma')|$ with

$$\kappa \geq 2^m \frac{(n-2)/2 - \sum_{j=1}^{n} d_j^{-1}}{mn[n]^m \left(1 + 2\sum_{z \in \mathcal{V}} (1 - d_z^{-1})\right)}.$$

$\square$

REMARK. The class of the graphs which are uniformly imbedded in the hyperbolic upper half–plane and satisfy a strong isoperimetric inequality is in fact much larger than the class just described: see e.g. Ancona (1988).

§7. Nonuniqueness and hyperbolic graphs

We turn now to the proof of the result of Cartwright and Woess. We start with some lemmas on the hyperbolic metric on Ω. Let us denote by $B(z, r)$ the open hyperbolic ball with centre $z \in \Omega$ and radius r. The hyperbolic area, still noted by $\sigma(r)$, of $B(z, r)$ does not depend on z. Writing $z = x + iy$, the expression of $\sigma(r)$ is

$$(4.53) \qquad \sigma(r) = \int_{B(z,r)} \frac{4}{(1 - |z|^2)^2} dx dy = 4\pi \sinh^2(r/2).$$

In this section $\Gamma = (V, Y)$ will always denote a graph which is uniformly embedded in Ω and satisfies a strong isoperimetric inequality. Examples of such graphs are provided (in the hyperbolic upper half–plane) in §6.

DEFINITION. The limit set of Γ is defined as the set V' of all the accumulation points of V on the unit circle (in the euclidean topology).

Lemma (4.54). *For every $w \in V$ we have*

(4.55) $$\deg(w) \leq \frac{\sigma(c + 1/2c)}{\sigma(1/2c)} \quad \text{where } c \text{ is as in (4.40).}$$

Furthermore

(4.56) $$\sum_{w \in V} (1 - |w|^2)^2 \leq \frac{4\pi e^{1/c}}{\sigma(1/2c)}.$$

Proof. If w_1 and w_2 are distinct vertices in V, then $d(w_1, w_2) \geq 1$, so that, by (4.40), $\rho(w_1, w_2) \geq c^{-1}$. Thus the balls $B(z, 1/2c)$, where z ranges among the neighbours of w, are mutually disjoint and contained in $B(w, c + 1/2)$. This implies immediately (4.55).

The hyperbolic ball $B(w, r)$ is a euclidean ball with centre on the ray from the origin trough w. This follows easily from the fact that the hyperbolic metric on Ω satisfies the equation

(4.57) $$\tanh(\rho(w_1, w_2)) = \frac{|w_1 - w_2|}{|1 - w_1 \overline{w}_2|} \quad \text{for every } w_1, w_2 \text{ in } \Omega.$$

Thus the minimum of $1 - |z|$ for z in the closure of $B(w, r)$ is attained at the point z_1 on this ray for which $|z_1| \geq |w|$ and $\rho(w, z_1) = r$. From (4.57) we have $\tanh(r/2) = (|z_1| - |w|)/(1 - |z_1||w|)$, whence $1 - |z_1| \geq (1 - |w|)e^{-r}$.

Let m denote the Lebesgue measure on the euclidean plane. Taking (4.53) into account, we obtain

$$\sigma(r) \leq m(B(w, r)) \frac{4}{(1 - |z_1|^2)^2}$$
$$= m(B(w, r)) \frac{4}{(1 - |z_1|)^2 (1 + |z_1|)^2}$$
$$\leq m(B(w, r)) \frac{4e^{2r}}{(1 - |w|^2)^2}.$$

Therefore we have for every $w \in V$

(4.58) $$m(B(w, r)) \geq (1 - |w|^2)^2 \frac{\sigma(r)}{4 \exp(2r)}.$$

Once more using the disjointness of the balls $B(w, 1/2c)$, $w \in V$, we obtain

(4.59) $$\pi = m(\Omega) \geq \sum_{w \in V} m(B(w, 1/2c)).$$

Combining (4.58) with (4.59), we obtain (4.56). $\square$

REMARK. Note that the arguments used to prove (4.55) show also that every compact subset of Ω must have finite intersection with V, so that indeed $V' \neq \emptyset$. This remark is important in view of the statement of Theorem (4.60) below.

DEFINITION. We say that a function ϕ on the euclidean closure $\overline{\Omega}$ of Ω satisfies the Lipschitz condition if there is a constant M such that, for every z_1, z_2 in $\overline{\Omega}$,

$$|\phi(z_1) - \phi(z_2)| \leq M|z_1 - z_2|.$$

Theorem (4.60). *Suppose that Γ is uniformly embedded in Ω and that Γ satisfies a strong isoperimetric inequality. Then, for every $\phi : \overline{\Omega} \mapsto \mathbb{C}$ which satisfies the Lipschitz condition there is a unique function h on $V \cup V'$ such that h coincides with ϕ on V' and the restriction of h to V is in $\mathbf{HD}$. In particular, if the limit set V' has more than one point, then $\mathbf{HD}$ contains nonconstant functions.*

Proof. Let f be the restriction of ϕ to V. Then

$$(4.61) \qquad D(f) \leq \frac{M^2}{2} \sum_{z_1 \sim z_2} |z_1 - z_2|^2.$$

If $z_1 \sim z_2$, then $\rho(z_1 - z_2) \leq c$, so that, by a well known hyperbolic inequality (see (7.2.4) in Beardon (1983)),

$$(4.62) \qquad |z_1 - z_2|^2 \leq (1 - |z_1|^2)(1 - |z_2|^2)\sinh^2(c/2).$$

By Cauchy–Schwarz inequality

$$(4.63) \qquad \begin{aligned} \sum_{z_1 \sim z_2} |z_1 - z_2|^2 &\leq \sinh^2(c/2)\Big(\sum_{z_1 \sim z_2} (|1 - z_1|^2)^2 \Big)^{1/2}\Big(\sum_{z_1 \sim z_2} (|1 - z_2|^2)^2 \Big)^{1/2} \\ &= \sinh^2(c/2) \sum_{z \in V} \deg(z)(|1 - z|^2)^2. \end{aligned}$$

Therefore $f \in \mathbf{D}$ by Lemma (4.54).

Let now $f = g + h$, with $g \in \mathbf{D}_0$ and $h \in \mathbf{HD}$, be the Royden decomposition of f. By (4.55) we may apply Theorem (4.27). Since Γ satisfies a strong isoperimetric inequality, the Sobolev–Dirichlet inequality (4.28) holds. Therefore g must vanish at infinity. It follows that h can be continuously extended to V' by setting $h(z) = \phi(z)$ for $z \in V'$.

Finally, if the limit set V' has more than one point, then there are functions ϕ satisfying the Lipschitz condition which are nonconstant on V'. The corresponding harmonic functions have nonconstant values on V' and hence, by continuity, are nonconstant on V. $\square$

REMARK. If Γ is a graph as in §6, then V' has cardinality greater than 1. Hence, in view of Theorem (4.60), the graphs studied in §6 provide examples of graphs with only one end and such that **HD** is nontrivial.

§8. Moderate growth and Foster's averaging formula

We turn now to the problem of extending Foster's averaging formula (Theorem (2.11)) to infinite graphs. In turn this will imply a uniqueness result much stronger than Corollary (4.13). The extension presented below is contained in the paper by Medolla and Soardi (1993). Suppose that (Γ, r) is an infinite network and let $x \sim y$ be adjacent vertices. In the following we will denote by $R_L(x,y)$ and $R_M(x,y)$ the limit and the minimal resistance, respectively, between x and y. By Corollaries (3.20) and (3.28)

$$(4.64) \qquad\qquad R_M(x,y) \le R_L(x,y).$$

Theorem (4.65). *Let (Γ, r) be a network such that the underlying graph Γ has moderate growth. Let $\{U_n\}$ be an increasing sequence of finite subsets of vertices such that $\cup_{n=0}^{\infty} U_n = V$ and $\lim_{n\to\infty} |\daleth(U_n)|/|U_n| = 0$. Then*

$$(4.66) \qquad \lim_{n\to\infty} \frac{1}{2|U_n|} \sum_{\substack{x\sim y \\ x,y\in U_n}} \frac{R_M(x,y)}{r(x,y)} = \lim_{n\to\infty} \frac{1}{2|U_n|} \sum_{\substack{x\sim y \\ x,y\in U_n}} \frac{R_L(x,y)}{r(x,y)} = 1.$$

Proof. For every n let $U_{1,n}, \ldots, U_{s_n,n}$ be the connected components of U_n. Let $\Gamma_{j,n} = (U_{j,n}, Y_{j,n})$ be the graphs induced by $U_{j,n}$ for all $j = 1, \ldots, s_n$. We have

$$U_n = \bigcup_{j=1}^{s_n} U_{j,n}, \quad \daleth(U_n) = \bigcup_{j=1}^{s_n} \daleth(U_{j,n})$$

and, if $i \ne j$,

$$\emptyset = U_{i,n} \bigcap U_{j,n}, \quad \emptyset = \daleth(U_{i,n}) \bigcap \daleth(U_{j,n}).$$

It follows that

$$(4.67) \qquad |U_n| = \sum_{j=1}^{s_n} |U_{j,n}|, \quad |\daleth(U_n)| = \sum_{j=1}^{s_n} |\daleth(U_{j,n})|.$$

Let $x, y \in U_n$, $x \sim y$. Then x and y belong to $U_{j,n}$ for some j. By Rayleigh's principle (Corollary (3.48)) the effective resistance $R_{j,n}(x,y)$ between x and y in the subnetwork $\Gamma_{j,n}$ is not smaller than $R_L(x,y)$. By Foster's formula and the first equation of (4.67) we have

$$\frac{1}{2|U_n|} \sum_{\substack{x \sim y \\ x,y \in U_n}} \frac{R_L(x,y)}{r(x,y)} \leq \frac{1}{2|U_n|} \sum_{j=1}^{s_n} \sum_{\substack{x \sim y \\ x,y \in U_{j,n}}} \frac{R_{j,n}(x,y)}{r(x,y)} \leq 1.$$

Thus we have

(4.68) $$\limsup_{n \to \infty} 2|U_n|^{-1} \sum_{\substack{x \sim y \\ x,y \in U_n}} \frac{R_L(x,y)}{r(x,y)} \leq 1.$$

Let $\tilde{U}_{j,n}$ denote the interior of $U_{j,n}$ i.e., the set of the vertices of $U_{j,n}$ which are not in $\daleth(U_{j,n})$. Let $(\Gamma'_{j,n}, r'_{j,n})$ be the network obtained by shorting together all the vertices of Γ not in $\tilde{U}_{j,n}$. Let $V'_{j,n} = \tilde{U}_{j,n} \cup \{b_{j,n}\}$ be the vertex set of $\Gamma'_{j,n}$.

For every $x, y \in V'_{j,n}$ such that $x \sim y$, denote by $R'_{j,n}(x,y)$ the effective resistance in $\Gamma'_{j,n}$ between x and y. Then, by Theorem (3.47) $R'_{j,n}(x,y) \leq R_M(x,y)$ if x and y are in the interior of $U_{j,n}$. Furthermore $R'_{j,n}(x, b_{j,n}) \leq R_M(x,y)$ for all $y \in \daleth(U_{j,n})$ and $x \in \tilde{U}_{j,n}$. If ϕ is as in (3.46) and $x \in V'_{j,n}$ is a neighbor of $b_{j,n}$, then

(4.69) $$\frac{1}{r'_{j,n}(x,b)} = \sum_{\phi(y)=b_{j,n}} \frac{1}{r(x,y)}.$$

Then we have, by Theorem (2.11) applied to $\Gamma'_{j,n}$

(4.70) $$\sum_{\substack{x \sim y \\ x,y \in V'_{j,n}}} \frac{R'_{j,n}(x,y)}{r'_{j,n}(x,y)} = 2|U_{j,n}| - 2|\daleth(U_{j,n})|.$$

Furthermore, by (4.69) and the preceding discussion

(4.71) $$\sum_{\substack{x \sim y \\ x,y \in V'_{j,n}}} \frac{R'_{j,n}(x,y)}{r'_{j,n}(x,y)} \leq \sum_{\substack{x \sim y \\ x,y \in U_{j,n}}} \frac{R_M(x,y)}{r(x,y)}.$$

Therefore, combining (4.67), (4.70) and (4.71) we get

$$\frac{1}{2|U_n|} \sum_{j=1}^{s_n} \sum_{\substack{x \sim y \\ x,y \in U_{j,n}}} \frac{R_M(x,y)}{r(x,y)} \geq \frac{1}{2|U_n|} \sum_{j=1}^{s_n} \sum_{\substack{x \sim y \\ x,y \in V'_{j,n}}} \frac{R'_{j,n}(x,y)}{r'_{j,n}(x,y)}$$

(4.72) $$= \frac{\sum_{j=1}^{s_n}(|U_{j,n}| - |\daleth(U_{j,n})|)}{|U_n|}$$

$$= 1 - \frac{|\daleth(U_n)|}{|U_n|}.$$

The thesis follows from (4.64), (4.68) and (4.72).   □

DEFINITION. Let (Γ, r) be a locally finite network. For every $x \in V$ the sums

$$R_M(x) = \sum_{y \sim x} R_M(x,y)$$

$$R_L(x) = \sum_{y \sim x} R_L(x,y)$$

are called the minimal and limit resistance degree at x, respectively. If Γ is vertex transitive then $R_M(x) = R_M$ and $R_L(x) = R_L$ do not depend on x.

Lemma (4.73). *Suppose that Γ is a vertex transitive graph of moderate growth, with all the resistances equal to 1. Let R_M and R_L be the minimal and the limit resistance degree at any vertex. Then $R_M = R_L = 2$. Consequently $R_M(x,y) = R_L(x,y)$ for any $x \sim y$.*

Proof. Let U_n be as in Theorem (4.65) and let (U_n, Y_n) be the (disconnected) graph induced by U_n. We have

$$(4.74) \qquad \frac{1}{2|U_n|} \sum_{\substack{x \sim y \\ x,y \in U_n}} R_M(x,y) = R_M \frac{|U_n| - |\daleth(Y_n)|}{2|U_n|},$$

where $\daleth(Y_n)$ denotes the set of all edges of Γ with exactly one endpoint in U_n. As $|\daleth Y_n| = O(|\daleth U_n|)$, letting $n \to \infty$ in (4.74) and remembering (4.66), we get $R_M = 2$. In the same way we see that $R_L = 2$.

If $R_L(x,y) > R_M(x,y)$ for some $x \sim y$ then $R_L > R_M$, which is absurd.   □

Thus we are lead to the following uniqueness result.

Theorem (4.75). *Suppose that Γ is a vertex transitive graph of moderate growth with all the resistances equal to 1. Then the network (Γ, r) belongs to $\mathcal{O}_{HD}$.*

Proof. For any dipole $\delta_y - \delta_x$, with $x \sim y$, let I_M and I_L be the minimal and limit currents, respectively, generated by the dipole. By Lemma (4.73), $\mathcal{W}(I_M) = \mathcal{W}(I_L)$. Therefore $I_M = I_L$, since the minimal current is unique by Theorem (3.21). Hence $(\Gamma, r) \in \mathcal{O}_{HD}$ by Theorem (4.4).   □

REMARK. Theorem (4.75) can be improved. We will prove in Chapter VII (Theorem (7.29)) that the conclusion of Theorem (4.75) holds for graphs of moderate growth with a finite number of orbits, a result already indicated by Doyle (1988).

REMARK. An example of vertex transitive graph with moderate but not poly-nomial growth was first exhibited by Grigorchuk (1983) and (1985). In fact Grig-orchuk constructed a finitely generated group whose Cayley graph has subexponen-tial but not polynomial growth. Further examples are contained in the paper by Fabrykowski and Gupta (1985).

As a consequence of Lemma (4.73) we can compute the effective minimal resis-tance across an edge of a 1–transitive graph (compare with Thomassen (1990)).

Corollary (4.76). *Let Γ be a locally finite 1–transitive graph all of whose edges have resistance 1. Then, the minimal effective resistance ρ across any edge of Γ is $2/d$, where d is the degree of any vertex of Γ.*

Proof. Assume first that the network is transient. Then, setting $f = d^{-1}(\delta_a - \delta_b)$ we have by symmetry

$$(4.77) \qquad \begin{aligned} \rho &= G(f)(a) - G(f)(b) = d^{-1}(G(a,a) - G(a,b) + G(b,b) - G(b,a)) \\ &= 2(d^{-1}(G(a,a) - G(a,b))). \end{aligned}$$

By symmetry

$$\rho = 2d^{-1}(G(a,a) - G(x,a)), \quad \text{where } x \text{ is any neighbor of } a.$$

Thus

$$\rho d = \frac{2}{d} \sum_{x \sim a} (G(a,a) - G(x,a)) = 2,$$

whence $\rho = 2/d$.

If the network is recurrent, then Γ cannot satisfy an isoperimetric inequality. Thus Γ has moderate growth by Theorem (4.39). The thesis follows from Lemma (4.73). $\square$

CHAPTER V

SOME EXAMPLES AND COMPUTATIONS

§1. Transience of infinite grids

The purpose of this chapter is to study in some detail certain aspects of remarkable classes of electrical networks such as infinite grids, cascades, trees and edge graphs of tilings of the plane. In some cases (especially in the case of grids) we will be able to carry out explicit computations of the solution of the discrete Poisson's equation

$$(5.1) \qquad (1 - P(u))(x) = u(x) - \sum_{y \in V} p(x,y)u(y) = f(x) \quad \text{for all } x \in V$$

where $f(x) = -c(x)^{-1}\imath(x)$ and $\imath \in \partial \mathbf{H}_1$. We start with the n–dimensional infinite grid $\mathbb{Z}^n$ (see Example (1.26)(1) for the definition).

Let $\Gamma = (V, Y)$ be a connected locally finite graph and let $o \in V$ be a reference vertex. A 1–chain I on Γ is called a flow of value m with single source at o if I is a flow on Γ generated by the 0–chain $-m\delta_o$ i.e., if

$$(5.2) \qquad \begin{aligned} \partial I(x) &= 0 \quad \text{if } x \neq o \\ \partial I(o) &= m. \end{aligned}$$

Let us introduce the notion of product of flows on certain graphs. Let $\Gamma = (V, Y)$ and $\Gamma' = (V', Y')$ be locally finite graphs and let $\mathbb{Z}$ the be 1–dimensional grid (or two–sided path). Let I and I' be 1–chains on $\Gamma \times \mathbb{Z}$ and on $\Gamma' \times \mathbb{Z}$ respectively, where $\times$ denotes the cartesian product of graphs (see (4.14)). The following definition of product of 1–chains is due to Lyons (1983).

DEFINITION. The Lyons product $J = I * I'$ of I and I' is the 1–chain on $\Gamma \times \Gamma' \times \mathbb{Z}$ defined as

$$(5.3) \quad j\left([(x,x',n),(x,x',n \pm 1)]\right) = \pm 2i\left([(x,n),(x,n \pm 1)]\right) i'\left([(x',n),(x',n \pm 1)]\right)$$

$$j\left([(x,x',n),(z,x',n)]\right) = i\left([(x,n),(z,n)]\right) i'\left([(x',n),(x',n+1)]\right) -$$
$$i\left([(x,n),(z,n)]\right) i'\left([(x',n),(x',n-1)]\right)$$

$$j\left([(x,x',n),(x,z',n)]\right) = i'\left([(x',n),(z',n)]\right) i\left([(x,n),(x,n+1)]\right) -$$
$$i'\left([(x',n),(z',n)]\right) i\left([(x,n),(x,n-1)]\right)$$

whenever $x \sim z$ or $x' \sim z'$. Note that J is actually a 1-chain.

The following lemma is an immediate consequence of the above definition (5.3).

Lemma (5.4). *Let I and I' be 1-chains on $\Gamma \times \mathbf{Z}$ and on $\Gamma' \times \mathbf{Z}$ respectively. Let $J = I * I'$ be their Lyons product. Then, for every $(x, x', n) \in V \times V' \times \mathbf{Z}$*

$$(5.5) \quad \partial J(x, x', n) = \partial I(x, n) \left(I' \left([(x', n), (x', n+1)] \right) - I' \left([(x, n), (x', n-1)] \right) \right) + \\ \partial I'(x', n) \left(I \left([(x, n), (x, n+1)] \right) - I \left([(x, n), (x, n-1)] \right) \right).$$

If I and I' are flows with single sources at $(o, 0)$ and $(o', 0)$ on $\Gamma \times \mathbf{Z}$ and on $\Gamma' \times \mathbf{Z}$ respectively, then $\partial J(x, x', n) = 0$ for all $n \neq 0$ and all $(x, x') \in V \times V'$. In the cases of interest it turns out that J is actually a flow on $\Gamma \times \Gamma' \times \mathbf{Z}$ with single source at $(o, o', 0)$.

For every $s = 1, 2, \ldots k$, let $x^{(s)}(n)$ be a positive nondecreasing integer valued function defined on the nonnegative integers. Assume that

$$(5.6) \quad \begin{aligned} x^{(s)}(0) &= 0 \qquad s = 1, \ldots, k \\ x^{(s)}(n+1) - x^{(s)}(n) &\leq 1 \quad \text{for every } s \text{ and } n. \end{aligned}$$

Let Υ be defined as the subgraph of $\mathbf{Z}^{k+1}$ induced by the subset of vertices

$$\left\{ (m_1, \ldots, m_k, n) \in \mathbf{Z}^{k+1} : n \geq 0, |m_s| \leq x^{(s)}(n) \text{ for } s = 1, \ldots, k \right\}.$$

Lyons proved the following characterization of the transience of Υ.

Theorem (5.7). *Suppose that every edge of Υ is assigned resistance 1. Then, the resulting network is transient if and only if*

$$(5.8) \quad \sum_{n=0}^{\infty} \frac{1}{\Pi_{s=1}^{k}(x^{(s)}(n) + 1)} < \infty.$$

Proof. If the series in (5.8) diverges, then the network is recurrent by the the Nash–Williams theorem (Theorem (3.50) in Chapter III). Actually, if we set for every nonnegative n

$$V_n = \{(x_1, \ldots, x_k, n) : |x_1| \leq x^{(1)}(n), \ldots, |x_k| \leq x^{(k)}(n)\},$$

then the series in (3.51) coincides with the series in (5.8).

To prove the converse implication it is sufficient, by Theorem (3.33), to construct a flow of finite energy on Υ with single source at the origin. In order to do this, let Υ_s be the subgraph of $\mathbf{Z}^2$ induced $\{(m, n) \in \mathbf{Z}^2 : n \geq 0, |m| \leq x^{(s)}(n)\}$. We define a 1–chain I_s on Υ_s in the following way. On a vertical edge $[(m,n),(m,n+1)]$, we set

$$i_s((m,n),(m,n+1)) = \frac{1}{2x^{(s)}(n)+1} \quad \text{for } |m| \le x^{(s)}(n).$$

On the horizontal edges $[(m,n),(m+1,n)]$ such that $x^{(s)}(n) = x^{(s)}(n-1)+1$ we set

$$i_s((m,n),(m+1,n)) = \frac{2m+1}{(2x^{(s)}(n)+1)(2x^{(s)}(n-1)+1)}$$
$$\text{for } 0 \le m \le x^{(s)}(n)-1,$$

$$i_s((m,n),(m-1,n)) = \frac{-2m+1}{(2x^{(s)}(n)+1)(2x^{(s)}(n-1)+1)}$$
$$\text{for } -x^{(s)}(n)+1 \le m \le 0.$$

We assign to I_s opposite values on the edges with opposite orientation. Finally, we set $i_s((m,n),(m+1,n)) = 0$ if $x^{(s)}(n) = x^{(s)}(n-1)$.

A simple computation shows that $\partial I_s(m,n) = 0$ if $(m,n) \ne (0,0)$ and $\partial I_s(0,0) = 1$. Hence I_s is a flow on Υ_s with source at $(0,0)$. Note also that

$$(5.9) \qquad |i_s((m,n),(m+1,n))| \le \frac{1}{2x^{(s)}(n)+1}.$$

Defining I_s to be zero on the edges of $\mathbb{Z}^2$ which are not edges of Υ_s, we may extend I_s to a flow on $\mathbb{Z}^2$ with the same source. Therefore the Lyons product $J = I_1 * I_2 * \cdots I_k$ of the flows I_s is well defined.

We see from (5.3) that J is a 1–chain on $\mathbb{Z}^{k+1}$ which vanishes on the edges of $\mathbb{Z}^{k+1}$ which are not edges of Υ. Thus J is a 1–chain on Υ. From (5.5) we have that J is in fact a flow on Υ with single source at the origin and value 2^{k-1}.

From (5.9) and (5.3) we see that on the edges of Υ we have

$$(5.10) \qquad \begin{aligned} |i((x,n),(x,n+1))| &\le \frac{2^k}{\Pi_{s=1}^k(2x^{(s)}(n)+1)} \\ |i((x,n),(x',n))| &\le \frac{C_k}{\Pi_{s=1}^k(2x^{(s)}(n)+1)} \end{aligned}$$

where C_k is a suitable constant. Set, for all $n \ge 0$,

$Y_1(n) = $ set of all edges of Υ of the form $[(x,n),(x,n+1)]$

$Y_2(n) = $ set of all edges of Υ of the form $[(x,n),(x',n)]$, with $x \sim x'$.

Then, $|Y_1(n)| = \Pi_{s=1}^k(2x^{(s)}(n)+1)$ and $|Y_2(n)| \le \Pi_{s=1}^k(2x^{(s)}(n)+1)$. Taking (5.10) into account we have for some positive constant c_k

$$\mathcal{W}(J) = \sum_{n=0}^\infty \Big(\sum_{B \in Y_1(n)} j^2(B) + \frac{1}{2} \sum_{B \in Y_2(n)} j^2(B) \Big)$$

$$\le c_k \sum_{n=0}^\infty \big(\Pi_{s=1}^k(x^{(s)}(n)+1) \big)^{-1}.$$

whence the thesis. □

The following result is a celebrated theorem of Polya (1921).

Theorem (5.11). Z^k *is transient if and only if* $k \geq 3$.

Proof. If $k \leq 2$, then Z^k is recurrent by Nash–Williams theorem. If $k \geq 3$, then we take in Theorem (5.7) $x^{(s)}(n) = n$ for every $s = 1, 2, \ldots, k$. Then Υ is transient, and Z^k is transient since it contains a transient subnetwork (Corollary (3.48)). □

Let Z_+ denote the one–ended infinite path represented by the nonnegative integers. Let Z_+^k denote the k–fold cartesian product of Z_+. All the resistances are equal to 1.

Corollary (5.12). *Let* $h \geq 0$, $k \geq 0$, $h + k > 0$. *Then,* $Z_+^k \times Z^h$ *is transient if and only if* $h + k \geq 3$.

Proof. Set $h + k = n$ and assume that $n \geq 3$. Suppose, by contradiction, that $Z_+^k \times Z^h$ is not transient. Denote by o the origin and by $\mathbf{P}_o^+$ the set of all one–ended infinite paths in $Z_+^k \times Z^h$ with first vertex o. Then, by Corollary (3.84), $\lambda(\mathbf{P}_o^+) = \infty$.

By (3) of Lemma (3.75) there exists a 1–chain I of finite energy on $Z_+^k \times Z^h$ such that $\sum_{B \in Y(\mathbf{p})} |i(B)| = \infty$ for all $\mathbf{p} \in \mathbf{P}_o^+$. We can extend I to a 1–chain J on the edges of Z^n by symmetry. Namely, let $x = (x_1, \ldots, x_k, \ldots, x_n)$ be any vertex in Z^n. We set

$$x^* = (|x_1|, |x_2|, \ldots, |x_k|, x_{k+1}, \ldots, x_n).$$

Clearly $x^* \in Z_+^k \times Z^h$. Then, for every edge $[x, y]$ of Z^n we set

$$J(x, y) = I(x^*, y^*).$$

Clearly J has finite energy and $\sum_{B \in Y(\mathbf{p})} |i(B)| = \infty$ for all $\mathbf{p} \in \mathbf{P}_o$. This implies that Z^n is recurrent, which is absurd.

Finally, for $h + k < 3$ recurrence follows at once from Theorem (5.7) and Corollary (3.48). □

REMARK. Another interesting case occurs when $k = 2$, $x^{(1)}(n) = n$, and $x^{(2)}(n)$ is the integer part of $(\log n)^\alpha$, for some $\alpha > 1$. In this case Theorem (5.7) implies that Υ is transient. Lyons describes such a graph Υ as obtained by a slight fattening of the quadrant in Z^2 (which is recurrent). Another way of fattening the quadrant in order to obtain a transient graph was described by Markvorsen, McGuinness and Thomassen (1992).

§2. Potentials in $\mathbb{Z}^n$

In this section we will compute the unique (by Theorem (4.17)) solution with finite energy of equation (5.1) in $\mathbb{Z}^n$. The problem of computing the electrical quantities related to $\mathbb{Z}^n$, especially in the case where the energizing 0–chain is a dipole and $n \leq 3$, is an important one in physics and in engineering. It appears several times in the literature: see e.g. Courant, Frederichs and Lewy (1928), McCrea and Whipple (1940) Stöhr (1950), Duffin (1953), Van der Pol (1959), Spitzer (1964), Flanders (1972), Zemanian (1984); see also the discussion in Flanders (1972).

The problem of studying equation (5.1) in the case of a semigrid both grounded and ungrounded was also studied; see Zemanian (1982) or Zemanian and Subramanian (1983). We will confine ourselves to $\mathbb{Z}^n$ and we will produce the explicit solution (in **D**) of Poisson's equation (5.1) for a general f. Such a solution will be expressed in terms of the potential kernel h of the simple random walk on $\mathbb{Z}^n$. Numerical estimates of such a kernel are beyond the scope of these notes. The asymptotics of h can be found in the literature cited above, at least in the cases $n = 2$ and $n = 3$.

Let e_j denote the j-th fundamental vector for $j = 1, 2, \ldots, n$ (all coordinates equal zero except the j-th which is equal to 1). The random walk operator P has the form

$$(5.13) \qquad P(u)(x) = (2n)^{-1} \sum_{j=1}^{n} \big(u(x + e_j) + u(x - e_j) \big).$$

The discrete Laplacian $(1 - P)$ is

$$(5.14) \qquad (1 - P)(u)(x) = -(2n)^{-1} \sum_{j=1}^{n} \big(u(x + e_j) + u(x - e_j) - 2u(x) \big).$$

The cardinality of the ball of radius m centered at the origin of $\mathbb{Z}^n$ has order m^n as $m \to \infty$, while the cardinality of the boundary of the same ball has order m^{n-1}. Hence $\mathbb{Z}^n$ does not satisfy an isoperimetric inequality and, by Corollary (4.27), we cannot solve the discrete Poisson's equation by inverting $(1 - P)$ on ℓ^2 (in fact $\mathbb{Z}^n$ satisfies a dimensional isoperimetric inequality; see Varopoulos (1985)).

Both the operators P and $1 - P$ are convolution operators on $\mathbb{Z}^n$. Recall that the convolution of two functions f and g on $\mathbb{Z}^n$ is defined as

$$f * g(x) = \sum_{y \in \mathbb{Z}^n} f(x - y)g(y)$$

whenever this expression makes sense. Set

$$(5.15) \qquad \chi = (2n)^{-1} \sum_{j=1}^{n} \big(\delta_{e_j} + \delta_{e_{-j}} \big),$$

where δ_x is the unit mass at x. Then (5.13) and (5.14) become respectively

$$P(u) = \chi * u, \quad (1 - P)(u) = (\delta_0 - \chi) * u.$$

Poisson's equation (5.1) becomes

(5.16) $$(\delta_0 - \chi) * u = f.$$

Denote by $\mathbb{T}^n$ the n–dimensional torus. For every integrable F on $\mathbb{T}^n$ the Fourier transform of F is the function $\mathcal{F}(F)$ on $\mathbb{Z}^n$ defined as

(5.17) $$\mathcal{F}(F)(x) = (2\pi)^{-n} \int_{-\pi}^{+\pi} \cdots \int_{-\pi}^{+\pi} F(\theta_1, \ldots, \theta_n) e^{-i(\theta_1 x_1 + \cdots + \theta_n x_n)} d\theta_1 \cdots d\theta_n$$

where $x = (x_1, \ldots, x_n)$.

If $\mathcal{F}(F) = f$, then we write $F = \mathcal{F}^{-1}(f)$. Clearly, $f \in \ell^2$ if and only if $F = \mathcal{F}^{-1}(f) \in L^2(\mathbb{T}^n)$.

A simple computation shows that

$$\mathcal{F}^{-1}(\delta_0 - \chi)(\theta_1, \ldots, \theta_n) = n^{-1}(2\sin^2 \theta_1/2 + \cdots + 2\sin^2 \theta_n/2).$$

Set, for $n \geq 3$,

(5.18) $$H(\theta_1, \ldots, \theta_n) = n(2\sin^2 \theta_1/2 + \cdots + 2\sin^2 \theta_n/2)^{-1}.$$

Clearly $H \in L^1(\mathbb{T}^n)$ if and only if $n \geq 3$.

DEFINITION. The function $h = \mathcal{F}(H)$ is called the potential kernel of the simple random walk on $\mathbb{Z}^n$, for $n \geq 3$. For $n = 2$ the potential kernel h is defined as

$$h(x_1, x_2) = \frac{1}{4\pi^2} \int_{-\pi}^{+\pi} \int_{-\pi}^{+\pi} \frac{1 - e^{-i(\theta_1 x_1 + \theta_2 x_2)}}{\sin^2 \theta_1/2 + \sin^2 \theta_2/2} d\theta_1 d\theta_2.$$

Theorem (5.19). *For every $n \geq 2$ the potential kernel h is a solution of Poisson's equation (5.1) with $f = \delta_0$. If $n \geq 3$, then $h(x - y) = G(x, y)$. If $n = 2$, then $h(x - y) - h(x) - h(y) = G(x, y; o)$, where o denotes the origin of the axes.*

Proof. If $n \geq 3$ the assertion follows immediately by taking the inverse Fourier transform $\mathcal{F}^{-1}$ of both sides of the equation

$$(\delta_0 - \chi) * u = \delta_0.$$

If $n = 2$, then

$$h(x \pm e_j) - h(x) = (2\pi)^{-2} \int_{-\pi}^{+\pi} \int_{-\pi}^{+\pi} \frac{(1 - e^{\mp i\theta_j})e^{-i(\theta_1 x_1 + \theta_2 x_2)}}{\sin^2 \theta_1/2 + \sin^2 \theta_2/2} d\theta_1 d\theta_2.$$

Summing these relations for $j = 1, 2$ and dividing by -4 we get

$$(1 - P)(h)(x) = (2\pi)^{-2} \int_{-\pi}^{+\pi} \int_{-\pi}^{+\pi} e^{-i(\theta_2 x_2 + \theta_2 x_2)} d\theta_1 d\theta_2$$

whence the thesis.

Suppose now $n \geq 3$. Set, for all $x, y \in \mathbf{Z}^n$ and every t, $0 < t \leq 1$,

$$g(x, y, t) = \sum_{k=0}^{\infty} t^k p^k(x, y) = \sum_{k=0}^{\infty} t^k \chi^{*k}(x - y)$$

where χ^{*k} is the k-th convolution power of χ.

Clearly g is well defined and $g(x, y, 1) = G(x, y)$. Moreover $g(x, y, t) \to G(x, y)$ as $t \to 1-$, by the monotone convergence theorem.

Now, setting $a(\theta_1, \ldots, \theta_n) = n^{-1} \sum_{j=1}^n \cos \theta_j$, we have

$$g(x, y, t) = \mathcal{F}((1 - ta(\theta_1, \ldots, \theta_n))^{-1})(x - y).$$

For every t, $0 < t \leq 1$ and every $\theta_1, \ldots, \theta_n$ we have

$$\frac{1}{1 - ta(\theta_1, \ldots, \theta_n)} \leq \frac{2}{1 - a(\theta_1, \ldots, \theta_n)}.$$

By the dominated convergence theorem $(1 - ta(\theta_1, \ldots, \theta_n))^{-1}$ converge in $L^1(\mathbf{T}^n)$ to $(1 - a(\theta_1, \ldots, \theta_n))^{-1}$ as $t \to 1$. As the L^1 convergence of the functions implies the uniform convergence of their Fourier transforms, $g(x, y, t)$ tends to $h(x - y)$ as $t \to 1$ for all $x, y \in \mathbf{Z}^n$.

Suppose now $n = 2$ and, as in Chapter III, let $G(x, y; o)$ denote the expected number of visits to y before hitting o, starting at x. We choose o to be $(0, 0)$, the origin of the axes.

By Theorem (3.38) $G(x, y; o)$ is the potential vanishing at o of the unique current with finite energy generated by the dipole $\imath = 4(\delta_o - \delta_y)$. On the other hand

$$h_y(x) = h(x - y) - h(x) = \mathcal{F}((e^{i(\theta_1 y_1 + \theta_2 y_2)} - 1)/(\sin^2 \theta_1/2 + \sin^2 \theta_2/2)).$$

Taking inverse Fourier transforms we see that

$$(\delta_0 - \chi) * h_y = \delta_y - \delta_0.$$

Therefore $h_y(x) = h(x - y) - h(x)$ is a potential of the current generated by the same dipole $\imath$. Furthermore, as $x \to \infty$, the following asymptotic formula for h is known (see Stöhr (1950) and Duffin and Shaffer (1960))

$$h(x) = 3\log 2 + 2\gamma + \frac{2}{\pi}\log|x|^2 + O(|x|^{-2})$$

where γ is the Euler constant and $|x|$ denotes the euclidean norm of x. Since the second differences of $\log|x|$ are $O(|x|^{-2})$ (compare with Flanders (1972)), it follows easily that for every fixed y there is a constant c (depending on y) such that

$$\sum_{x \in \mathbb{Z}^2} |h_y(x \pm e_j) - h_y(x)|^2 \le c \sum_{z \ne 0} |x|^{-2} < \infty \quad \text{for } j = 1, 2.$$

Hence $h_y \in \mathbf{D}$, so that $h_y(x) = G(x, y; o) + \kappa$ for some κ depending on y. Since $h(o) = 0$, we have $\kappa = h(-y) = h(y)$. □

Set, for every function u on $\mathbb{Z}^n$ and $j, k = 1, 2, \ldots n$

$$\Delta_j(u)(x) = u(x - e_j) - u(x) = u * (\delta_{e_j} - \delta_0)(x),$$
$$\Delta^2_{k,j}(u)(x) = \Delta_k(\Delta_j(u))(x).$$

By (5.17) and the definition of h, we have that $\Delta_j(h)$ is the Fourier transform of the function

$$n(e^{i\theta_j} - 1)(2\sin^2\theta_1/2 + \cdots + 2\sin^2\theta_n/2)^{-1}$$

which is square integrable for all $n \ge 3$.

On the other hand, $\Delta^2_{k,j}(h)$ is the Fourier transform of the function

$$(5.20) \qquad F_{k,j}(\theta_1, \ldots, \theta_n) = \frac{n(e^{-i(\theta_k + \theta_j)} - e^{-i\theta_k} - e^{-i\theta_j} + 1)}{(2\sin^2\theta_1/2 + \cdots + 2\sin^2\theta_n/2)}$$

which is bounded on $\mathbb{T}^n$ for all $n \ge 2$.

Let us return to Poisson's equation. A necessary condition in order that (5.1) admits a solution $u \in \mathbf{D}$ is that $-2nf$ is the boundary of a 1-chain in $\mathbf{H}_1$. It follows that f must be of the form

$$f = \sum_{j=1}^n \sum_{z \in \mathbb{Z}^n} a_{j,z}(\delta_{x+e_j} - \delta_x) \quad \text{with} \quad \sum_{j=1}^n \sum_{z \in \mathbb{Z}^n} a^2_{j,z} < \infty.$$

Setting $f_j = \sum_{z \in \mathbb{Z}^n} a_{j,z}\delta_z$ (note that f_j belongs to $\ell^2(\mathbb{Z}^n)$) we have

$$(5.21) \qquad f = \sum_{j=1}^n (\delta_{e_j} - \delta_0) * f_j.$$

Theorem (5.22). *A necessary and sufficient condition for the discrete Poisson equation on $\mathbb{Z}^n$ to admit a solution $u \in \mathbf{D}$ is that f is of the form (5.21), with $f_j \in \ell^2(\mathbb{Z}^n)$ for all $j = 1, \ldots, n$. Such a solution u is unique up to an additive constant. We have for $n \geq 3$*

$$u(x) = \sum_{j=1}^{n} \Delta_j(h) * f_j,$$

while for $n = 2$

(5.23) $$u(x) = \sum_{j=1}^{2} \sum_{y \in \mathbb{Z}^2} \left(\Delta_j(h)(x - y) - \Delta_j h(y) \right) f_j(y).$$

Proof. The convolution in $\Delta_j(h) * f_j$ is well defined for $n \geq 3$, since in this case $f_j \in \ell^2$ and $\Delta_j(h) \in \ell^2$. If $n = 2$, we see that for every fixed x $\Delta_j(h)(x - y) - \Delta_j(h)(y)$ is the value at y the Fourier transform of

$$(e^{i(\theta_1 x_1 + \theta_2 x_2)} - 1)(e^{i\theta_j} - 1)(\sin^2 \theta_1 / 2 + \sin^2 \theta_2 / 2)^{-1}$$

which is bounded on $\mathbb{T}^2$. Therefore $\Delta_j(h)(x - y) - \Delta_j(h)(y)$ is in $\ell^2(\mathbb{Z}^2)$ and the series in (5.23) converges absolutely.

Fix an edge $[x - e_k, x]$. Then, for every $n \geq 2$,

$$u(x - e_k) - u(x) = \sum_{j=1}^{n} f_j * \Delta^2_{k,j}(h)(x).$$

By (5.20) the product $\mathcal{F}^{-1}(f_j) F_{k,j}$ is in $L^2(\mathbb{T}^n)$ so that $f_j * \Delta^2_{k,j}(h)$ is in $\ell^2(\mathbb{Z}^n)$. It follows that $D(u) < \infty$. Finally

$$(\delta_0 - \chi) * u = \sum_{j=1}^{n} (\Delta_j(h) - \chi * \Delta_j(h)) * f_j = \sum_{j=1}^{n} \Delta_j(h - \chi * h) * f_j$$

$$= \sum_{j=1}^{n} \Delta_j(\delta_0) * f_j = \sum_{j=1}^{n} (\delta_{e_j} - \delta_0) * f_j$$

$$= f$$

so that u is the solution of the discrete Poisson's equation (5.1). $\square$

REMARK. According to Theorem (3.30) if $n \geq 3$ and $\sum 2nG(|f|)(x)|f(x)| < \infty$, then $-2nf \in \partial \mathbf{H}_1$ and the potential u has the form $u(x) = G(f)(x)$ (up to an additive constant). By Theorem (5.19), $u = h * f$. Note that this expression coincides with the potential found in Theorem (5.22). (compare also with Lemma (1.32)).

Analogously, if $n = 2$ and $\sum_x c4G(|f|; o)(x)|f(x)| < \infty$, then, by Theorem (3.38), $-2nf \in \partial\mathbf{H}_1$ and $u(x) = G(f; o)(x)$ is the potential which is 0 at the origin. By Theorem (5.19) again, $u(x) = \sum_y (h(x - y) - h(y) - h(x))f(y)$ coincides, up to an additive constant, with the potential (5.23).

§3. Ungrounded cascades

Grounded and ungrounded cascades are widely studied infinite electrical networks (see the bibliography in Zemanian (1991)). We will be concerned with currents and potentials generated by dipoles in a cascade. The exposition in this section and in the next one is mainly based on Zemanian's book (1991) and on the thesis of Affer (1992).

Suppose we are assigned a sequence of finite networks (Γ_j, r_j), for $j = 0, 1, \ldots$, $\Gamma_j = (V_j, Y_j)$ with the following property: for $i \neq j$, Γ_i and Γ_j have no common edges and, if $|i - j| > 1$, no common vertices. For every j there exist four distinct vertices $x_j^1, x_j^2, y_j^1, y_j^2$ in V_j such that

$$(5.24) \qquad x_j^2 = x_{j+1}^1, \quad y_j^2 = y_{j+1}^1 \quad \text{for all } j = 0, 1, \ldots.$$

Furthermore Γ_j and Γ_{j+1} do not have any other common vertex.

We can define an infinite network (Γ, r) as union of the networks (Γ_j, r_j). Namely, we set

$$(5.25) \qquad V = \bigcup_{j=0}^{\infty} V_j, \quad Y = \bigcup_{j=0}^{\infty} Y_j.$$

The resistance $r(B)$ of an edge $B \in Y$ is defined in the following way. Let j be the unique subscript such that $B \in Y_j$. We set

$$(5.26) \qquad r(B) = r_j(B).$$

DEFINITION. The network (Γ, r), where $\Gamma = (V, Y)$ is defined by (5.25) and r by (5.26), is called a one–ended ungrounded cascade (see Figure 5.1). If the subscript j ranges from $-\infty$ to $+\infty$ and all the other assumptions hold true, then the network (Γ, r) defined as in (5.25), (5.26) (with j ranging in $(-\infty, +\infty)$) is called a two–ended ungrounded cascade.

The networks (Γ_j, r_j) appearing in this chainlike structure are called stages of the cascade (and also two–ports networks). The common nodes in Γ_j and Γ_{j+1} are

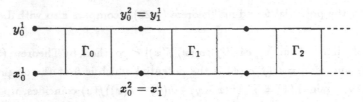

$$y_0^2 = y_1^1$$

Fig. 5.1

called terminals. If $j = 0$, then the nodes x_0^1 and y_0^1, which are not connected to any other stage, will be simply denoted by x_0 and y_0.

In the sequel we will deal mainly with one–ended cascades. The corresponding results for two–ended cascades can be obtained with obvious modifications.

DEFINITION. A one–ended (resp. two–ended) ungrounded cascade is called uniform if all the stages are copies of the same finite network (Υ, ρ), where $\Upsilon = (V(\Upsilon), Y(\Upsilon))$. By this we mean that there exist distinct vertices x^1, x^2, y^1, y^2 in $V(\Upsilon)$ and, for every j, an isomorphism $\phi_j : V(\Upsilon) \mapsto V_j$ such that $\phi_j(x^i) = x_j^i$, $\phi_j(y^i) = y_j^i$, for $i = 1,2$ and $j = 0,1,2,\dots$ (resp. $-\infty < j < +\infty$). Moreover $r_j(\phi(x), \phi(y)) = \rho(x,y)$ for all $x \sim y$ in $V(\Upsilon)$.

EXAMPLE (5.27). The two–ended lattice is the uniform two–ended ungrounded cascade where all the resistances are 1 and every stage is a copy of the graph Υ with vertex and edge sets defined in the following way

$$V(\Upsilon) = \{x^1, x^2, y^1, y^2\},$$
$$Y(\Upsilon) = \{[x^k, x^h], [y^k, y^h], [x^k, y^h], [y^h, x^k], \ h \neq k, h, k = 1, 2\}.$$

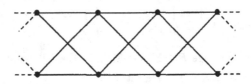

Fig. 5.2

Theorem (5.28). Let (Γ, r) be a one–ended ungrounded cascade. Let

(5.29)
$$\alpha_j = \sum_{z \in V_{j+1}} c(x_j^2, z), \quad \beta_j = \sum_{z \in V_{j+1}} c(y_j^2, z).$$

If

(5.30)
$$\sum_{j=0}^{\infty} (\alpha_j + \beta_j)^{-1} = \infty,$$

then the cascade is recurrent.

Proof. The theorem is a straightforward consequence of the Nash–Williams Theorem (3.50), where the sets of the canonical decomposition are precisely the vertex sets V_j of the stages Γ_j minus the points x_j^2 and y_j^2.

Corollary (5.31). *Every one–ended ungrounded uniform cascade is recurrent. Hence every $\imath \in \partial H_1$ generates a unique current with finite energy.*

REMARK. It is easy to see that a theorem similar to Theorem (5.28) holds for two–ended cascades.

Now we will show how to compute the effective resistance between the two first terminals x_0 and y_0 of an ungrounded one–ended uniform cascade. First we note the following reciprocity principle for recurrent networks (see Zemanian (1991) and Tetali (1991) in the finite case).

Lemma (5.32). *Let (Γ, r) be a recurrent network, and denote by a_k, b_k, $k = 0, 1$, two couples of nodes such that $a_k \neq b_k$. Let $\imath_k = \delta_{b_k} - \delta_{a_k}$ and denote by u_k the potential of the minimal current generated by $\imath_k$, for $k = 0, 1$. Then*

$$u_0(a_1) - u_0(b_1) = u_1(a_0) - u_1(b_0).$$

Proof. Let us assume $u_k(b_0) = 0$. Then, by (3) of Theorem (3.38),

$$u_0(x) = c(a_0)^{-1} G(x, a_0; b_0), \quad u_1(x) = ca_1^{-1} G(x, a_1; b_0) - c(b_1)^{-1} G(x, b_1; b_0).$$

Now, $c(x)G(x, y; b_0) = c(y)G(y, x; b_0)$ whenever x and y are different from b_0. Hence

$$
\begin{aligned}
u_0(a_1) - u_0(b_1) &= c(a_0)^{-1}(G(a_1, a_0; b_0) - G(b_1, a_0; b_0)) \\
&= c(a_1)^{-1} G(a_0, a_1; b_0) - c(b_1)^{-1} G(a_0, b_1; b_0) \\
&= u_1(a_0) - u_1(b_0).
\end{aligned}
$$

$\square$

Lemma (5.33). *Let (Γ, r) be a recurrent network and let a_k and b_k be as in Lemma (5.32). Set $\imath_k = i_k(\delta_{b_k} - \delta_{a_k})$ and let u_k be the potential of the corresponding currents. Let u be the potential of the current generated by $\jmath = \imath_0 - \imath_1$. Then*

(5.34)
$$
\begin{aligned}
u(a_0) - u(b_0) &= R_{00} i_0 - R_{01} i_1 \\
u(a_1) - u(b_1) &= R_{10} i_0 - R_{11} i_1
\end{aligned}
$$

where $i_h R_{hk} = u_h(a_k) - u_h(b_k)$ (i.e., R_{hk} is the effective resistance between a_k and b_k due to the current generated by $\imath_h$). Furthermore $R_{01} = R_{10}$.

Proof. Let J be the current generated by $\jmath$, and let I_h be the current generated by $\imath_h$ ($h = 0, 1$). Then $J = I_2 - I_1$ and $u = u_0 - u_1$. Therefore $u(a_0) - u(b_0) =$

$u_0(a_0) - u_1(b_0) = R_{00}i_0 - R_{01}i_1$. The same argument proves the second equation of (5.34). The equality $R_{01} = R_{10}$ is a consequence of Lemma (5.32). □

The numbers R_{hk} appearing in the above Lemma are also called the open–circuit resistance parameters of Υ (once that a_k and b_k have been fixed). Note that $R_{00} > 0$, $R_{11} > 0$ and $R_{01} = R_{10} \geq 0$.

We are now ready to study effective resistances, voltages and currents in uniform cascades. Let (Γ, r) be a uniform one–ended ungrounded cascade. Let $\imath_0 = i_0(\delta_{y_0} - \delta_{x_0})$, $i_0 \neq 0$. Denote by I the unique (by Theorem (5.29)) current generated by $\imath_0$ in the cascade. and let ∂_n be the boundary operator restricted to the graph $\Gamma_0 \cup \Gamma_1 \cup \cdots \cup \Gamma_n$. Then, $\partial_n I(x) = 0$ for all the nodes $x \in V_0 \cup \cdots \cup V_n$, different from the terminals x_0, y_0, x_n^2, y_n^2. Since (see Lemma (2.1))

$$\sum_{x \in V_0 \cup \cdots \cup V_n} \partial_n I(x) = 0, \quad \text{and} \quad \partial_n I(x_0) = -\partial_n I(y_0) = -i_0,$$

we have

(5.35) $$\partial_n I(x_n^2) = -\partial_n I(y_n^2).$$

Set $i_{n+1} = -\partial_n I(x_n^2)$, $n \geq 0$. Since $\partial I(x) = 0$ (for x different from x_0 and y_0), by (5.35) the current entering the uniform cascade $\Gamma_{n+1} \cup \Gamma_{n+2} \cup \cdots$ is i_{n+1} at x_{n+1}^1 and $-i_{n+1}$ at y_{n+1}^1.

Theorem (5.36). *Let (Γ, r) be a uniform one–ended ungrounded cascade. Let i_n be as above and let R_{00}, R_{11} and $R_{01} = R_{10}$ be the open–circuit resistance parameters of Υ. Then, the effective resistance R between x_0 and y_0 is*

(5.37) $$R = \frac{R_{00} - R_{11} + \sqrt{(R_{00} + R_{11})^2 - 4R_{01}^2}}{2}.$$

If u denotes the potential of the current I generated by $\imath_0 = i_0(\delta_{y_0} - \delta_{x_0})$, then

(5.38) $$\begin{aligned} i_n &= i_0(\alpha - \sqrt{\alpha^2 - 1})^n, \\ u(x_n^1) - u(y_n^1) &= R i_0(\alpha - \sqrt{\alpha^2 - 1})^n \end{aligned}$$

where $\alpha = (R_{00} + R_{11})/2R_{10}$.

Proof. By the definition of i_n, the restriction of u to the vertices of the one–ended uniform cascade whose graph is $C_{n+1} = \Gamma_{n+1} \cup \Gamma_{n+2} \cup \cdots$ is the potential of the current generated in this cascade by $\imath_{n+1} = i_{n+1}(\delta_{y_{n+1}^1} - \delta_{x_{n+1}^1})$. By uniformity, the resistance between x_{n+1}^1 and y_{n+1}^1 in C_{n+1} is exactly R. This implies that

(5.39) $$u(x_n^1) - u(y_n^1) = R i_{n+1} \quad \text{for } n = 0, 1, \ldots.$$

Analogously, the restriction of u to the vertices of Γ_n is the potential of the current generated by $i_n - i_{n+1}$ in (Γ_n, r).

Let us consider the case $n = 0$. We apply Lemma (5.33). In this case, by (5.39), $u(x_0) - u(y_0) = Ri_0$ and $u(x_0^2) - u(y_0^2) = u(x_1^1) - u(y_1^1) = Ri_1$. Equations (5.34) become

(5.40)
$$Ri_0 = R_{00}i_0 - R_{01}i_1$$
$$Ri_1 = R_{10}i_0 - R_{11}i_1.$$

The homogeneous system (5.40) has a nontrivial solution (i_0, i_1) if and only if

(5.41) $$R^2 - R(R_{00} - R_{11}) + R_{10}^2 - R_{00}R_{11} = 0.$$

Therefore R is the only positive root of equation (5.40), whence (5.37).

To compute the ratio i_n/i_{n+1}, we proceed as follows. By Lemma (5.33) applied to Γ_n and by (5.39) we have

(5.42)
$$Ri_n = u(x_n^1) - u(y_n^1) = R_{00}i_n - R_{01}i_{n+1}$$
$$Ri_{n+1} = u(x_n^2) - u(y_n^2) = R_{10}i_n - R_{11}i_{n+1}.$$

Taking quotients in equations (5.42) and then solving with respect to i_n/i_{n+1} we get

$$\frac{i_n}{i_{n+1}} = \frac{R_{00} + R_{11} - \sqrt{(R_{00} + R_{11})^2 - 4R_{10}^2}}{2R_{10}}.$$

Expressions (5.38) follow immediately. $\square$

REMARK. The proof of Theorem (5.36) above is based on Theorem 6.1–4 in Zemanian (1991). As a consequence of the proof (existence of a unique positive solution of equation (5.31)) we have that $R_{01}^2 \leq R_{00}R_{11}$. In fact strict inequality holds: see Zemanian (1982).

§4. Grounded cascades. Generalized networks

Suppose that, as in the definition of ungrounded cascades, we have a sequence of finite networks (Γ_j, r_j), for $j = 0, 1, \ldots$, $\Gamma_j = (V_j, Y_j)$. We require the following conditions: there exists a unique vertex (the ground) o such that $o \in V_j$ for all j. For $i \neq j$, Γ_i and Γ_j have no common edges and, if $|i - j| > 1$, no common vertices besides o. For every j there exist two distinct vertices x_j, y_j in V_j such that

$$x_j \neq o, \quad y_j \neq o, \quad y_j = x_{j+1}$$

for all $j = 0, 1, \dots$. Moreover, Γ_j and Γ_{j+1} do not have any other common vertex. Finally we require that the total conductance at o is finite, i.e.

$$(5.43) \qquad \sum_{j=0}^{\infty} \sum_{\substack{x \in V_j \\ x \sim o}} r_j^{-1}(x, o) < \infty$$

Let Γ and r be defined as in (5.25) and (5.26) respectively. Note that Γ is locally infinite at o: see Figure 5.3.

DEFINITION. The network (Γ, r) just defined is called a one–ended grounded cascade with stages (Γ_j, r_j) and terminals x_j, y_j, o.

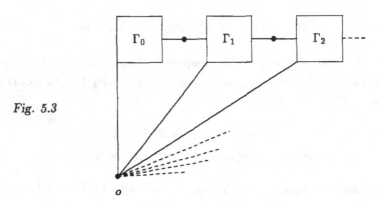

Fig. 5.3

Two–ended ungrounded cascade are defined in the similar way, assuming that the stages are indexed by j ranging from $-\infty$ to $+\infty$.

The following theorem is due to Affer (1992).

Theorem (5.44). *A grounded one–ended cascade (Γ, r) does not admit any nonconstant harmonic function.*

Proof. Assume that u is a real nonconstant harmonic function on Γ. There exist two vertices a_0 and a_1, $a_0 \sim a_1$, such that $u(a_0) < u(a_1)$. Since $u(a_1) = \sum_{x \sim a_1} p(a_1, y) u(y)$, there exists $a_2 \sim a_1$ such that $u(a_1) < u(a_2)$. Analogously, there exists $b_{-1} \sim a_0$ such that $u(b_{-1}) < u(a_0)$. Hence there exist two one–ended infinite paths $\mathbf{p}^+$ and $\mathbf{p}^-$, with vertices $a_0 \sim a_1 \sim a_2 \sim \cdots$ and $b_{-1} \sim b_{-2} \sim b_{-3} \sim \cdots$ respectively, such that $u(a_j) < u(a_{j+1})$, if $j \geq 0$ and $u(b_j) < u(b_{j+1}) < u(a_0)$ if $j < 0$. In particular the two paths do not have common vertices.

Now, every V_j is finite, and Γ_j and Γ_{j+1} have in common only the terminals $y_j = x_{j+1}$ and o. Since the cascade is one ended, all but finitely many terminals must be vertices of both paths $\mathbf{p}^+$ and $\mathbf{p}^-$, which is absurd. $\square$

Suppose now that, as in the case of ungrounded cascades, we try to define the notion of a uniform grounded cascade, i.e. a cascade in which all stages are copies of the same network (Υ, ρ). We see immediately that condition (5.43) cannot be satisfied, since

$$\sum_{j=0}^{\infty} \sum_{\substack{x \in V_j(\Upsilon) \\ x \sim o}} \rho^{-1}(x, o) = \infty.$$

Therefore these networks do not fall within the scope of the theory outlined in the previous chapters. Existence theorems for currents in networks having nodes of infinite total conductance were studied by Zemanian: see Zemanian (1991)(a) and the references therein included. Here we take a simpler approach in the spirit of Theorem (3.21).

DEFINITION. Let Γ be countable graph and let $r : Y \mapsto \mathbb{R}_+$ a positive function such that $r(x, y) = r(y, x)$ for all $x \sim y$. The couple (Γ, r) is called a generalized electrical network.

Note that we do not require that (1.1) holds. The nodes x where (1.1) holds are called finite. Otherwise we say that the node x is infinite.

The definition of the boundary of a 1–chain at an infinite node is problematical. Nevertheless we may define the cycles. Namely, let C be the space of all 1–chains I which belongs to $\mathbf{H}_1$ and have the following property: if x is an infinite node, then $i(x, y) = 0$ for all but finitely many neighbors $y \sim x$. Clearly C contains all the finite 1–chains, so that C is never trivial, not even if all the nodes are infinite. In particular, the boundary operator is well defined on C.

Let Z be the space of all 1–chains $Z \in C$ such that $\partial Z(x) = \sum_{y \sim x} z(x, y) = 0$.

DEFINITION. A 1–chain $Z \in \mathbf{H}_1$ is called a cycle if it belongs to the closure Z^* of Z in $\mathbf{H}_1$.

DEFINITION. We say that a 1–chain $I \in Z^*$ is a current generated by $E \in \mathbf{H}_1$ if

(5.45) $\qquad (I - E, Z) = 0 \qquad$ for all finite cycle Z

The following theorem is the analogue of Theorem (3.21). Since the proof is the same it will be omitted.

Theorem (5.46). *Let $\tilde{Z}$ be any closed subspace of Z^* containing all the finite cycles. For every $E \in \mathbf{H}_1$ there exists a unique 1–chain $\tilde{I} \in \tilde{Z}$ satisfying (5.45) for all $Z \in \tilde{Z}$. Such a 1–chain is the orthogonal projection of E onto the space $\tilde{Z}$. Furthermore, if $\tilde{Z}_1 \subseteq \tilde{Z}_2$, then we have, for the corresponding currents $\tilde{I}_1$ and $\tilde{I}_2$,*

$$W(\tilde{I}_2 - E) \leq W(\tilde{I}_1 - E).$$

It is clear that in the case of generalized networks we may still define the potential of a current as the unique (up to a constant) function g on V such that $g(x) -$

$g(y) = r(x,y)(i(x,y) - e(x,y))$ for every edge $[x,y]$. We have $D(g) = \mathcal{W}(I - E)$.
Furthermore, for every finite node x we define, as usual, $p(x,y) = c(x)^{-1}c(x,y)$ and,
for any function f and every finite node x, we set $P(g)(x) = \sum_{y \sim x} p(x,y)g(y)$.

If g is the potential of a current I generated by E, then we have

$$(5.47) \qquad (I - P)(g)(x) = -c(x)^{-1}\partial E(x) \quad \text{for every finite node } x.$$

In general, there is no full analogue of Theorem (3.13): the transition probabilities $p(x,y)$ might be undefined for every x. It is likely that generalized networks
are related to finitely additive random walks on graphs; see e.g. Kuhn (1991).

Assume for the moment that (Γ, r) is a generalized network with only one infinite
vertex o. Then $P(g)$ is defined at every node except o. If I is a current in $\mathbf{H}_1$
generated by E, then its potential g satisfies (5.47) for all $x \neq o$. Clearly, $D(g) < \infty$.
Conversely if g is a solution of (5.47) with $D(g) < \infty$, then the 1–chain I such that
$i(x,y) = c(x,y)(g(x) - g(y)) + e(x,y)$ satisfies (5.45) and belongs to $\mathbf{H}_1$. We do
not know whether, in general, $I \in \mathbf{Z}^*$. We will see below that this is the case
for uniform grounded cascades. In any case we can say that there exists a unique
current generated by $E \in \mathbf{H}_1$ if there are no nonconstant functions f such that
$D(f) < \infty$ and $P(f)(x) = f(x)$ for all $x \neq o$.

DEFINITION. A uniform one–ended, resp. two–ended, grounded cascade is defined as a one–ended, resp. two–ended, grounded cascade such that all the stages
are copies of a single finite network (Υ, ρ) (with $\Upsilon = (V(\Upsilon), Y(\Upsilon))$).

By this we mean that there exist three distinct vertices $\overline{x}, \overline{y}, \overline{o}$ in $V(\Upsilon)$ and, for
every j, an isomorphism $\xi_j : V(\Upsilon) \mapsto V_j$ such that

$$\xi_j(\overline{o}) = o, \quad \xi_j(\overline{x}) = x_j, \quad \xi_j(\overline{y}) = y_j$$

for all $j = 0, 1, 2, \ldots$, resp. $-\infty < j < +\infty$. Moreover $r_j(\xi_j(a), \xi_j(b)) = \rho(a,b)$ for
all $a \sim b$ in $V(\Upsilon)$.

Grounded uniform cascades are examples of generalized networks with only one
infinite vertex. In this case we can show that there are no nonconstant harmonic
functions on $V \setminus \{o\}$ with finite Dirichlet sums.

Theorem (5.48). *Let* (Γ, r) *be either any two–ended uniform grounded cascade
or a one–ended uniform grounded cascade with the property that there exists an
automorphism* ψ *of the network* (Υ, ρ) *such that* $\psi(\overline{x}) = \overline{y}$, $\psi(\overline{y}) = \overline{x}$, $\psi(\overline{0}) = \overline{o}$. *If*
u *is a function on* V *such that* $D(u) < \infty$ *and*

$$u(x) = \sum_{y \sim x} p(x,y)u(y) \quad \text{for all } x \neq o,$$

then u is constant.

Proof. Let first (Γ, r) be two–ended. We define an automorphism $\phi : V \mapsto V$ in the following way: for every relative integer j and every $z \in V_j$, we set $\phi(z) = \xi_{j+1}(\xi_j^{-1}(z))$. In other words we map z into the "same" vertex of the next stage (Γ_{j+1}, r_{j+1}). In particular $\phi(o) = o$.

Clearly ϕ is an automorphism of the graph Γ and $r(\phi(z_1), \phi(z_2)) = r(z_1, z_2)$. It follows that the function $u(\phi(x))$ is harmonic at every point $z \in V$, $z \neq o$, and that the same is true for the function $f(x) = u(\phi(x)) - u(x)$. In particular $f(o) = 0$.

Since $D(u) < \infty$, we have that $D(f) < \infty$. Then, by an argument similar to the proof of Theorem (4.8), $f(x) \to 0$, as $x \to \infty$. Thus, for every $\epsilon > 0$ there is n such that, for all $|j| \geq n$ and all $x \in V_j$, $|f(x)| < \epsilon$. Let $U = \cup j \leq nV_j$. By Theorem (1.35), $|f(x)| < \epsilon$ for all $x \in U$. It follows that $f(x) = 0$ for all $x \in V$.

Thus $u(x) = u(\phi(x))$. But then

$$\sum_{x \sim y \in V_j} c(x, y)|u(x) - u(y)|^2 = \sum_{x \sim y \in V_{j+1}} c(x, y)|u(x) - u(y)|^2,$$

so that $D(u) = \infty$, unless u is constant.

Assume now that the cascade is one–ended. The cascade (Γ, r) can be seen as "half" of a two–ended cascade $(\Gamma^*, r*)$, where $\Gamma^* = \cup_{j=-\infty}^{+\infty} \Gamma_j$ each stage (Γ_j, r_j) being isomorphic to the same network (Υ, ρ).

Any function u harmonic on (Γ, r) can be extended to a function u^* harmonic on (Γ^*, r^*). Namely, if $x \in V_j$ with $j \geq 0$, then we set $u^*(x) = u(x)$. If $x \in V_{-j-1}$ for some $j \geq 0$, then let $z \in V_j$ be such that $z = \xi_j \circ \psi \circ \xi_{-j-1}^{-1}(x)$. We set $u^*(x) = u(z)$.

The function u^* is well defined since $\psi(o) = o$ and

$$x_0 = \xi_0 \circ \psi \circ \xi_{-1}^{-1}(y_{-1}) \quad \text{and}$$
$$x_j = \xi_j \circ \psi \circ \xi_{-j-1}^{-1}(y_{-j-1}), \quad y_{j-1} = \xi_{j-1} \circ \psi \circ \xi_{-j}^{-1}(x_{-j}) \quad \text{for } j \geq 1.$$

Clearly we have to verify the harmonicity of u^* only at the points $x_{-j} = y_{-j-1}$, for $j = 0, 1 \ldots$. We have

$$(5.49) \qquad \sum_{\substack{z \sim x_0 \\ z \in V_0}} c(x_0, z)(u(z) - u(x_0)) = 0,$$

since u is harmonic in (Γ, r). Furthermore, for every $x \in V_{-1}$, $x \sim y_{-1} = x_0$, we have by definition $u^*(x) = u(z)$, with $z \sim x_0$ and $c(y_{-1}, x) = c(x_0, z)$ (since ψ is a network isomorphism). By (5.49)

$$\sum_{\substack{x \sim y_{-1} \\ x \in V_{-1}}} c(y_{-1}, x)(u^*(x) - u(y_{-1})) = \sum_{\substack{z \sim x_0 \\ z \in V_0}} c(x_0, z)(u(z) - u(x_0)) = 0.$$

Therefore u is harmonic at x_0. An analogous argument shows that u^* is harmonic at all the other junction points $x_j = y_{j-1}$.

Finally, $D(u) < \infty$ implies $D(u^*) < \infty$. By the first part of the theorem we may conclude that $u^* = \text{const.}$, so that u is constant as well. $\square$

Corollary (5.50). *Let (Γ, r) be a uniform grounded cascade satisfying the same assumptions as in Theorem (5.38). Let $E \in \mathbf{H}_1$. Then there exists a unique current of finite energy generated by E.*

Proof. A current exists by Theorem (5.46). Since there are no nonconstant functions f such that $D(f) < \infty$ and $(1 - P)(f)(x) = 0$ for all $x \neq o$, the current is unique. $\square$

EXAMPLE (5.51). The condition on one–ended grounded uniform cascades in Theorem (5.48) is far from being necessary. For instance Let Υ be the graph

$$V(\Upsilon) = \{\bar{x}, \bar{y}, \bar{o}\},$$

$$Y(\Upsilon) = \{[\bar{x}, \bar{y}], [\bar{y}, \bar{x}], [\bar{x}, \bar{o}], [\bar{o}, \bar{x}]\}.$$

Set $\rho(\bar{x}, \bar{y}) = q^{-1}$, $\rho(\bar{x}, \bar{o}) = p^{-1}$. Let (Γ, r) be the one–ended grounded uniform cascade with stages isomorphic to (Υ, ρ) see Figure 5.4.

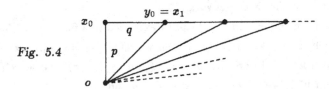

Fig. 5.4

A function u such that $u(o) = 0$ is harmonic on $V/\{o\}$ if and only if

$$(p + q)u(x_0) = qu(x_1),$$

$$qu(x_{j+1}) - (2q + p)u(x_j) + qu(y_{j-1}) = 0 \quad j = 1, 2, \ldots.$$

Therefore u is harmonic on $V/\{o\}$ if and only if

$$u(x_j) = A\left(\frac{2q + p + \sqrt{p^2 + 4pq}}{2q}\right)^j + B\left(\frac{2q + p - \sqrt{p^2 + 4pq}}{2q}\right)^j$$

where A and B are constants such that

$$A(-p + \sqrt{p^2 + 4pq}) = B(p + \sqrt{p^2 + 4pq}).$$

Then, $D(u) < \infty$ if and only if $A = B = 0$. Clearly, there is no automorphism of the network (Υ, r) exchanging $\bar{x}$ and $\bar{y}$ and leaving $\bar{o}$ fixed.

§5. The strong isoperimetric inequality for trees

Throughout this section T will denote an infinite locally finite tree with no vertices with degree 1 (the so-called terminals). An unbranched path $\mathbf{p}$ in T is defined as a path whose vertices have degree 2.

We will prove below Gerl's necessary and sufficient condition for a tree to satisfy an isoperimetric inequality (Gerl (1986)). The proof of the following lemma is essentially the same as in Brooks (1985).

Lemma (5.52). *Let T be a tree without terminals and suppose that $d = \min \deg(x) \geq 3$. Then T satisfies a strong isoperimetric inequality with constant $(d-2)/(d-1)$.*

Proof. Assume, by contradiction, that there exists a nonempty finite set L such that

$$(5.53) \qquad |\daleth(L)| < \frac{d-2}{d-1}|L|.$$

Among all the finite sets satisfying (5.53), we choose a set L with the smallest cardinality. We fix a reference vertex $o \in L$ and introduce the following notation: $|x| = d(x,o)$ and, for every finite subset K and every $w \in K$,

$$K(w) = \{z \in K : z \sim w, |z| = |w| + 1\}.$$

Let $z \in L$ be such that $|z|$ is maximum in L. Let w be the father of z i.e., the unique vertex of T such that $|w| = |z| - 1$ and $w \sim z$.

If $w \notin L$, then $z \in \daleth(L)$ and, defining L' as $L \setminus \{z\}$, we see that L' satisfies (5.53) as well. Assume now that $w \in L$. If $w \in \daleth L$, then we set $L' = L \setminus L(w)$. Clearly $w \in \daleth(L')$, and we easily get

$$\frac{|\daleth(L')|}{|L'|} = \frac{|\daleth(L)| - |L(w)|}{|L| - |L(w)|} < \frac{|\daleth(L)|}{|L|} < \frac{d-2}{d-1}.$$

Finally, suppose that w is in L but not in $\daleth(L)$. Then $L(w) = \{t \in T : t \sim w, |t| = |w| + 1\}$, so that $|L(w)| \geq d-1$. As $|\daleth(L)| \leq |L| - 1$, we get by (5.53)

$$\frac{|\daleth(L')|}{|L'|} = \frac{|\daleth(L)| - |L(w)| + 1}{|L| - |L(w)|} \leq \frac{|\daleth(L)| - d + 2}{|L| - d + 1} < \frac{|\daleth(L)|}{|L|} < \frac{d-2}{d-1}.$$

In any case L' is a proper subset of L which still satisfies (5.53). This contradicts our choice of L. $\square$

REMARK. If T is homogeneous of degree d, then it is not hard to prove that $(d-2)/(d-1)$ is the best constant.

A version of this Lemma for trees having arbitrary resistances was proved by Cattaneo (1993).

Theorem (5.54). *An infinite tree T without terminals satisfies a strong isoperimetric inequality if and only if there exists a constant κ such that the length of any unbranched path in T is smaller than κ.*

Proof. If there are arbitrarily long unbranched paths, then T cannot satisfy a strong isoperimetric inequality.

Conversely, assume that L is a finite subset of vertices of T. We modify the tree T by replacing every unbranched path $x_0 \sim x_1 \sim \cdots \sim x_n$ by a single edge $[x_0, x_n]$. We denote by T' and L' the new tree and the new set thus obtained. We have $\kappa|L'| \geq |L|$ and $|\partial L| \geq |\partial L'|$. By Lemma (5.52) we get

$$|\partial L| \geq |\partial L'| \geq \frac{1}{2}|L'| \geq \frac{1}{2\kappa}|L|.$$

This concludes the proof. $\square$

EXAMPLE (5.55). Assume that T is homogeneous of degree d. Then, T is a Cayley graph of the free product $G = \mathbb{Z} * \cdots * \mathbb{Z} * \mathbb{Z}_2 * \cdots * \mathbb{Z}_2$ of k copies of $\mathbb{Z}$ and h copies of the two–element group $\mathbb{Z}_2$ ($h + 2k = d$). Namely, let $a_1^{\pm 1}, \ldots, a_k^{\pm 1}$ be the generators of the k copies of $\mathbb{Z}$ and let $b_1, \ldots, b_h$ be the generators of the h copies of $\mathbb{Z}_2$. The homogeneous tree T is the Cayley graph of G with respect to these generators.

It is immediately seen that, for every function on G, $P(f)(x) = f * \chi(x)$, where

$$\chi = d^{-1} \sum_{j=1}^{k} (\delta_{a_j} + \delta_{a_j^{-1}}) + d^{-1} \sum_{j=1}^{h} \delta_{b_j},$$

and the convolution $f * g$ is given

$$f * g(x) = \sum_{y \in G} f(y^{-1}x)g(y)$$

whenever the above expression is defined for all x.

By Corollary (4.37) $\partial \mathbf{H}_1 = \ell^2$. We have $G(x,y) = \sum_{n=0}^{\infty} \chi^{*n}(y^{-1}x) = (1 - \chi)^{-1}(y^{-1}x)$. In particular, for every $f \in \ell^2$,

$$\sum_{y} c(x)G(x,y)|f(x)||f(y)| = (G(|f|), |f|)$$

$$\leq \|G\|\|f\|_2^2,$$

where $\|G\|$ denotes the norm of G as an operator on ℓ^2; see Theorem (4.27). Hence the potential of the minimal current generated by f is $G(f)$ (Theorem (3.30)).

§6. Capacity of the boundary of a tree

This section is devoted Benjamini and Peres' characterization of transient trees (Benjamini and Peres (1992)). We start with some definitions.

Let (T, r) be an infinite network whose underlying graph T is a locally finite tree with no vertices of degree 1. Choose a reference vertex o. We assume that for all infinite paths $o = x_0 \sim x_1 \sim x_2 \sim \cdots$ starting from o we have

$$(5.56) \qquad \sum_{j=1}^{\infty} r(x_{j-1}, x_j) = \infty.$$

Denote by T^* the set of all finite or infinite paths starting from o. Clearly, there is a one–to–one correspondence between the vertex set of T and the set of all finite paths starting from o, so that we may identify T with a subset of T^* (we are using the same notation T for the graph and its vertex set, since no confusion can arise). The elements in $bT = T^* \setminus T$ can be viewed as the "ends" of the infinite geodesics from o. In fact it is possible to show that the set of the ends introduced in §4 of Chapter IV can be identified with bT.

DEFINITION. Let ξ and η be elements of T^*. The confluent (with respect to o) of two paths $\xi, \eta \in T^*$ is defined as the common vertex z for which $d(z, o)$ is maximum. Let $o = x_0 \sim \cdots \sim x_n = z$ be the path (identified with z) joining the confluent itself with o. We set

$$(\xi|\eta) = \sum_{j=1}^{n} r(x_j, x_{j-1}).$$

The Gromov distance $\rho(\xi, \eta)$ between ξ and η is defined as

$$(5.57) \qquad \rho(\xi, \eta) = e^{-(\xi|\eta)} \quad \text{if } \xi \neq \eta, \quad \rho(\xi, \xi) = 0.$$

It is not hard to verify that T^* endowed with the Gromov metric is a compact metric space, with dense discrete subset T. Namely, suppose that ξ_j is sequence in T^*. If the distances $d(\xi_j, o)$ are uniformly bounded, then there is a constant subsequence. Assume now that the distances $d(\xi_j, o)$ are not uniformly bounded or undefined. Then, for every n there is a subsequence $\xi_{n,j}$ and a vertex y such that $d(y, o) = n$ and such that y belongs to every $\xi_{n,j}$. The standard diagonal argument and (5.56) allow us to pick a convergent subsequence in the Gromov metric.

The compact set $bT = T^*/T$ is also called the "boundary" of T (see Cartier (1972)).

DEFINITION. Let (M, ρ) be a compact metric space. For every finite Borel measure μ on M we set

$$\mathcal{I}(\mu) = -\int_M \int_M \log \rho(\xi, \eta) d\mu(\xi) d\mu(\eta)$$

$$= \int_M \phi_\mu(\eta) d\mu(\eta),$$

where

(5.58) $\phi_\mu(\eta) = -\int_M \log \rho(\xi, \eta) d\mu(\xi)$ for all $\eta \in M$.

The function ϕ_μ in (5.58) is called the logarithmic potential of μ. The logarithmic capacity $\text{cap}(B)$ of a Borel subset $B \subseteq M$ is defined as

$$\text{cap}(B) = \sup\{\mu(B) : \mu \text{ positive with } \mathcal{I}(\mu) \leq 1\}.$$

We study the case where $M = bT$ and ρ is the Gromov distance. In this case

$$\phi_\mu(\eta) = \int_{bT} (\xi|\eta) d\mu(\xi).$$

Lemma (5.59). *The network (T, r) is transient if and only if $\text{cap}(bT) > 0$.*

Proof. Suppose $\text{cap}(bT) > 0$. Then, there exists a positive measure ν on bT such that $\mathcal{I}(\nu) \leq 1$. For every $x \neq o$ let $C(x)$ be the "cone"

(5.60) $C(x) = \{\xi \in bT : \text{the geodesic from } o \text{ to } \xi \text{ passes through } x\}$

and let $1_{C(x)}$ be its indicator function. We will also use the notation introduced in §2 of Chapter II: for any two distinct vertices x and y we write $x \succ y$ if x is a vertex of the unique path from o to y. The set of all $y \sim x$ such that $x \succ y$ will be denoted by $U(x)$.

Let $\eta \in bT$ be the infinite geodesic $o = x_0 \sim x_1 \sim \cdots$. Then

(5.61)
$$\phi_\mu(\eta) = \int_{bT} (\xi|\eta) d\mu(\xi) = \int_{bT} \left(\sum_{j=1}^{\infty} 1_{C(x_j)}(\xi) r(x_{j-1}, x_j) \right) d\nu(\xi)$$

$$= \sum_{j=1}^{\infty} \nu(C(x_j)) r(x_{j-1}, x_j).$$

We extend ϕ_ν to the whole of T^* in accordance with (5.61). If $o = x_0 \sim \cdots \sim x_n = x$ is the geodesic from o to the vertex x, we define

(5.62) $\phi_\nu(x) = \sum_{j=1}^{n} \nu(C(x_j)) r(x_{j-1}, x_j), \qquad \phi_\nu(o) = 0.$

It is easy to see that ϕ_ν is continuous on T^*.

For every $y \in U(x)$ we have from (5.62)

$$(5.63) \qquad \phi_\nu(y) - \phi_\nu(x) = \nu(C(y))r(x,y).$$

By (5.63) $(1-P)(\phi_\nu)(o) = -c(o)^{-1}\nu(bT)$ and $(1-P)(\phi_\nu)(x) = 0$ for $x \neq o$. Hence ϕ_ν is subharmonic. Now we claim that $\mathcal{I}(\nu) = D(\phi_\nu)$.

By (5.63), we have

$$(5.64) \quad \frac{1}{2}\sum_{\substack{|x|\leq n \\ |y|\leq n}} c(x,y)|\phi_\nu(x) - \phi_\nu(y)|^2 = \sum_{|x|\leq n}\phi(x)\sum_{y\in V(x)} c(x,y)(\phi_\nu(x) - \phi_\nu(y))$$

$$+ \sum_{|x|=n}\sum_{y\in U(x)} c(x,y)\phi_\nu(x)(\phi_\nu(x) - \phi_\nu(y)) = \sum_{|x|=n}\phi_\nu(x)\nu(C(x)).$$

Let f_n be the function on bT defined as

$$f_n = \sum_{|x|=n}\phi_\nu(x)1_{C(x)}.$$

By (5.61) and (5.62) the sequence $\{f_n(\eta)\}$ is increasing with n for every $\eta \in bT$ and $f_n(\eta) \to \phi_\nu(\eta)$ as $n \to \infty$. By Levi's theorem

$$\sum_{|x|=n}\phi_\nu(x)\nu(C(x)) = \int_{bT} f_n(\eta)d\nu(\eta) \to \mathcal{I}(\nu).$$

Thus $\mathcal{I}(\nu) = D(\phi_\nu)$ by (5.64). Since ϕ_ν is not constant, Theorem (3.34) implies that (T,r) is transient.

Conversely, suppose that (T,r) is transient. By Corollary (3.32) the function $G(x,o)$ is harmonic except at o (in fact is superharmonic) and has finite Dirichlet sum. It is clear that $G(x,o) \geq G(y,o)$ for every $x \succ y$. The function $\phi(x) = G(o,o) - G(x,o)$ is harmonic except at o, nonnegative and has finite Dirichlet sum. We may define a nontrivial positive measure ν on the cones (5.60) by setting

$$(5.65) \qquad \nu(C(y)) = c(x,y)(\phi(y) - \phi(x)) \quad \text{for all } x \neq o \text{ and all } y \in U(x).$$

(compare with (5.63)). As these cones form a basis of closed and open sets for the topology of bT, (5.65) defines ν uniquely as a positive bounded Borel measure on bT. Note that the additivity of ν follows from the harmonicity (except at o) of ϕ.

By the definition of ϕ, $\lim \phi(x)$ exists and is finite as $x \to \eta$, for every $\eta \in bT$. Therefore we may extend ϕ to a continuous function on T^*. By (5.61), (5.62) and (5.63) we see that $\phi = \phi_\nu$. Hence ϕ_ν is bounded and $\text{cap}(bT) > 0$. $\square$

REMARK. The above proof is obtained by a slight modification of Benjamini and Peres' original argument. It would be interesting to prove a similar result for other classes of networks.

Theorem (5.66). *Let T be an infinite locally finite tree without terminals. Let r satisfy (5.56). Then the following are equivalent*

(1) *The network (T, r) is transient.*

(2) *There exists a constant κ such that for every n there exist n distinct vertices $x_1, \ldots, x_n$ in T such that*

$$\binom{n}{2}^{-1} \sum_{i<j} (x_i | x_j) < \kappa.$$

(3) *There exists a constant κ' such that for any finite set of vertices $x_1, \ldots, x_n$ in T there is a vertex $z \in T$ which does not lie on the geodesic between o and $x_1, \ldots, x_n$ and yet satisfies*

$$\frac{1}{n} \sum_{i=1}^{n} (z | x_i) < \kappa'.$$

Proof. For any compact metric space (X, ρ) define the generalized diameters

$$\delta_n(X) = \inf_{x_1, \ldots, x_n} \binom{n}{2}^{-1} \sum_{i<j} \log \frac{1}{\rho(x_i, x_j)}$$

and the Chebyshev constants

$$M_n(X) = \sup_{x_1, \ldots, x_n} \inf_y \frac{1}{n} \sum_{j=1}^{n} \log \frac{1}{\rho(y, x_j)}.$$

A classical theorem by Fekete and Szegö asserts that the sequence of the generalized diameters is increasing and that

(5.67) $$\lim_{n \to \infty} \delta_n(X) = \lim_{n \to \infty} M_n(X) = \operatorname{cap}(X)^{-1}$$

(see Carleson (1967), Theorem 6 on page 37). Taking $X = bT$ and ρ the Gromov metric we have

$$\delta_n(bT) = \inf_{x_1, \ldots, x_n} \binom{n}{2}^{-1} \sum_{i<j} (x_j | x_j)$$

$$M_n(bT) = \sup_{x_1, \ldots, x_n} \inf_y \frac{1}{n} \sum_{j=1}^{n} (y | x_j).$$

Suppose that the network is transient. By Lemma (5.59) $\operatorname{cap}(bT) > 0$ so that (5.67) implies that the sequences $\delta_n(bT)$ and $M_n(bT)$ are bounded. For each n we can select $\xi_1, \ldots, \xi_n$ in bT such that, for some κ, $\binom{n}{2}^{-1} \sum_{i<j} (\xi_i | \xi_j) < \kappa$. Choose a vertex x_i on the geodesic from o to ξ_i in such a way that the x_i are pairwise distinct. This proves that (1) implies (2). In the same way one proves that (1) implies (3).

Assume now that (2) holds. We want to replace the vertices x_i in (2) by boundary points $\xi_i \in bT$ in such a way that

$$(5.68) \qquad\qquad (x_i | x_j) = (\xi_i | \xi_j).$$

This in not possible, in general, for the original tree T. Thus, we enlarge T by adding a new geodesic ray (a one–ended path) above every vertex of T. We assign resistance e.g. equal to 1 to every edge of the ray. This creates a new tree $\hat{T}$ such that (5.68) can be achieved on $b\hat{T}$ (when x_i are vertices of T). Hence $\delta_n(b\hat{T})$ is a bounded sequence, so that $\operatorname{cap}(b\hat{T}) > 0$.

We see from (5.61) and (5.56) that $\mathcal{I}(\nu) = \infty$ for every positive discrete measure ν. Since $b\hat{T}/bT$ is countable, it follows that $\operatorname{cap}(bT) > 0$. Hence (2) implies (1). In the same way one shows that (3) implies (1).

EXAMPLE (5.69). Following R. Lyons (1993) and Benjamini and Peres (1992) we construct by induction a recurrent tree with exponential growth. The root o is the only vertex in level 0 and such a root is connected with two vertices in level 1. For $n \geq 1$ and $1 \leq k \leq 2^{n-1}$, the k-th vertex in level n has three neighbors, numbered $3k-2$, $3k-1$, $3k$, in level $n+1$. For $2^{n-1}+1 \leq k \leq 2^n$ the k-th vertex has a unique neighbor numbered $k + 2^n$ in level $n+1$. A moment's thought shows that, with only one exception, all the one–sided paths ξ starting at o have the following property: all but finitely many vertices of ξ have only one neighbor in the next level. It follows that bT is countable (it suffices to map each boundary point to the first vertex on its path which has only one neighbor in the next level). Assign resistance 1 to every edge. Then $\operatorname{cap}(bT) = 0$ and the tree is recurrent by Lemma (5.59). It is also clear that T has exponential growth, since it has 2^n vertices in level n.

§7. Edge graphs of tilings of the plane

DEFINITION. Let A be an unbounded, connected and simply connected subset of the euclidean plane $\mathbb{R}^2$. A tiling of A is a family $\mathcal{T}$ of closed topological disks $T \subseteq A$ (the tiles) having pairwise disjoint interiors and such that $\bigcup_{T \in \mathcal{T}} T = A$. Any connected component of the intersection of two or more tiles is called a vertex or an edge of the tiling, depending on whether it is a single point or an arc. Unless otherwise stated, from now on V and Y will denote the set of vertices and the set of edges of the tiling. To avoid pathologies arising from singular points (see e.g. Breen (1985)), we shall always assume that T is locally finite i.e., every closed disk intersects only finitely many tiles. Under this assumption the boundary of every tile is a finite union of edges. Each edge y of T is a connected component of the intersection of exactly two tiles, T and some $T' \in \mathcal{T}$. No point of y other than its endpoints belongs to any tile different from T and T'. A point is a vertex if and only if it is the connected component of the intersection of at least three tiles i.e., if and only if it is the endpoint of an edge. We refer to the book by Grünbaum and Shephard (1987) for the proof of these facts as well as for hundreds of examples and illustrations.

DEFINITION. The plane graph $\Gamma = (V, Y)$ is called the edge graph of $\mathcal{T}$.

Note that the edge graph of $\mathcal{T}$ is at most countable and connected with no self loops. Moreover each vertex has finite degree at least 3. To rule out multiple edges between two vertices we will always assume that every tile has at least three edges.

In the sequel we will always assume that the family $\mathcal{T}$ is countable and we will identify the infinite graph Γ with its plane representation.

DEFINITION. An infinite locally finite tiling $\mathcal{T}$ of A will be called quasi normal if the following condition are satisfied

(1) there exists n such that every tile T has at most n edges;

(2) there exists δ such that $\operatorname{diam} T < \delta$ for every tile T;

(3) there exists κ such that, for every tile T, $R_T \leq \kappa r_T$,

where r_T is the radius of the largest circle contained in T and R_T is the radius of the smallest circle containing T.

All the normal tilings in the sense of Grünbaum and Shephard are quasi normal tilings. The class of quasi normal tilings is however much larger, and it contains tilings of exponential growth.

EXAMPLE (5.70). Let us describe a quasi normal tiling of the first quadrant A of the euclidean plane whose edge graph has exponential growth.

The vertices of the edge graph Γ are the points $z \in A$ such that

$$z = (k + m2^{-k}, 2 + (n-2)2^{-k}) \quad \text{for all } m, n, k = 0, 1, 2 \ldots$$

The edges of Γ are the line segments connecting the points $(k + m2^{-k}, 2 + (n-2)2^{-k})$, $(k + (m+1)2^{-k}, 2 + (n-2)2^{-k})$, and the points $(k + m2^{-k}, 2 + (n-2)2^{-k})$, $(k + m2^{-k}, 2 + (n-1)2^{-k})$. It is clear that Γ is the edge graph of a quasi normal tiling of A; see Figure 5.5.

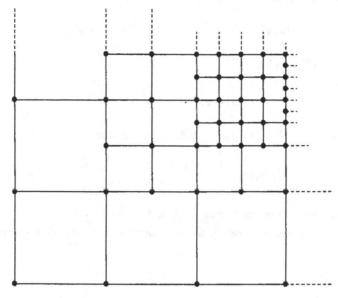

Fig. 5.5

For $q = 1, 2, \ldots$ let V_q denote the subset of V whose elements are:

$(q, 0)$

$(q, 1), (q + 1/2, 1)$

$(q, 1 + 1/2), (q + 1/4, 1 + 1/2), (q + 1/2, 1 + 1/2), (q + 1/2 + 1/4, 1 + 1/2)$

$\cdots\cdots\cdots\cdots\cdots\cdots\cdots\cdots\cdots\cdots$

$(q, 2 - 2^{-q}), \ldots, (q + m2^{-q}, 2 - 2^{-q}), \ldots, (q + 1 - 2^{-q}, 2 - 2^{-q}).$

Set $z_q = (q, 0)$. We have $d(x, z_q) \leq 2q + 1$ for all $x \in V_q$ and $|V_q| = 1 + 2 + \cdots + 2^q = 2^{q+1} - 1$. Denoting by B_n be the ball of radius n centered at the origin,

$$\bigcup_{q=1}^{n} V_q \subseteq B_{3n+1} \quad \text{for all } n = 1, 2, \ldots.$$

Since the sets V_q are pairwise disjoint, $|B_{3n}| \geq 2^{n+2} - n - 3$. Hence Γ has exponential growth.

Let us assign resistance 1 to every edge of a quasi normal tiling of A. The following result was proved in Soardi (1990).

Theorem (5.71). *Let Γ be the edge graph of a quasi normal tiling T of $\mathbb{R}^2$. Then Γ is recurrent.*

Proof. For every $(x,y) \in \mathbb{R}^2$ let

$$\alpha(x,y) = (10\delta)^{-1} \log(1 + \log(1 + x^2 + y^2)),$$

where δ is as in (2) and let

$$f(x,y) = \sin \alpha(x,y).$$

We set $\rho_k^2 = \exp(e^{10k\delta\pi} - 1) - 1$, $k = 1, 2, \ldots$ and define

$$f_k(x,y) = \begin{cases} f(x,y) & \text{if } x^2 + y^2 \leq \rho_k^2 \\ 0 & \text{if } x^2 + y^2 \geq \rho_k^2. \end{cases}$$

Let ϕ and ϕ_k denote the restriction of f and f_k to V. Note that ϕ_k is finitely supported, since the tiling is locally finite. The proof consists of two steps.

1st step.

We claim that

$$(5.72) \qquad\qquad D(\phi - \phi_k) \to 0 \quad \text{as} \quad k \to \infty$$

so that $\phi \in \mathbf{D}_0$.

Set $\psi_k = \phi - \phi_k$ and let B_k denote the closed disk of radius ρ_k centered at the origin. Then

$$(5.73) \qquad\qquad 2D(\psi_k) = \sum_{T \in \mathcal{T}_k} \sum_{\substack{z \sim w \\ z,w \in V(T)}} |\psi_k(z) - \psi_k(w)|^2,$$

where $V(T)$ denotes the set of all vertices of the tile T and $\mathcal{T}_k \subseteq \mathcal{T}$ is the set af all tiles having at least one vertex outside B_k.

So, let $z \sim w$ and $z, w \in V(T)$ for some $T \in \mathcal{T}_k$. Suppose first that both z and w are not in B_k. Then, denoting by $|\cdot|$ the euclidean norm,

$$(5.74) \quad |\psi_k(z) - \psi_k(w)|^2 \leq$$

$$|z - w|^2 \int_0^1 \left[f_x^2((1-\theta)z + \theta w) + f_y^2((1-\theta)z + \theta w) \right] d\theta \leq$$

$$|z - w|^2 \int_0^1 h((1-\theta)z + \theta w) d\theta,$$

where

$$h(x,y) = (\delta\rho)^{-2} \log^{-2}(1 + \rho^2) \quad \text{with } \rho = \rho(x,y) = (x^2 + y^2)^{1/2}.$$

There is a positive constant c such that, for large values of ρ, and for every ρ^* such that $|\rho - \rho^*| \leq 2\delta$

$$(5.75) \qquad \frac{h(\rho)}{h(\rho^*)} \leq c.$$

Hence, if $z_T^* \in T$ is such that $h(z_T^*) = \min_{z \in T} h(z)$, then $|z_T^* - (1 - \theta)z - \theta w)| \leq 2\delta$ (for $0 \leq \theta \leq 1$), by assumption (2). Thus, by (5.74) and (5.75)

$$(5.76) \qquad |\psi_k(z) - \psi_k(w)|^2 \leq c|z - w|^2 \min_{z \in T} h(z).$$

If $w \in B_k$ and $z \notin B_k$, then let w^* denote the intersection of the boundary of B_k with the line segment joining z and w. Then, $\psi_k(w) = \psi_k(w^*) = 0$ and, as before,

$$|\psi_k(z) - \psi_k(w)|^2 = |\psi_k(z) - \psi_k(w^*)|^2$$
$$\leq |z - w|^2 \int_0^1 h((1 - \theta)z + \theta w^*)d\theta.$$

The same argument as above now gives (5.76) once again. Hence (5.76) holds in any case. Now, by assumption (3)

$$(5.77) \qquad |z - w|^2 \leq 4R_T^2 \leq 4\kappa^2 r_T^2 \leq 4\kappa^2 \pi^{-1} \text{meas}(T),$$

where $\text{meas}(T)$ denotes the Lebesgue measure of T. Since by (1) there are at most n edges for each tile, combining (5.73), (5.76) and (5.77) we get

$$D(\psi_k) \leq 2cn\kappa^2 \pi^{-1} \sum_{T \in \mathcal{T}_k} \min_{z \in T} h(z) \, \text{meas}(T)$$
$$\leq \text{const.} \int_{x^2+y^2>(\rho_k-2\delta)^2} h(x,y) \, dx \, dy.$$

It follows that $D(\psi_k)$ tends to 0 as $k \to \infty$. Hence $\phi \in \mathbf{D_0}$.

2nd step.

Now we shall prove that, for any one–sided infinite path $\mathbf{p}$ in Γ, starting at any vertex $o \in V$, $\phi(z)$ does not have limit as z tends to infinity along the vertices of $\mathbf{p}$.

Any such path is an infinite non self–intersecting curve obtained by the union of infinitely many edges $[z_n, z_{n+1}]$, where $z_n \in V$ and $o \sim z_1 \sim z_2 \sim \ldots \sim z_n \sim \ldots$ are the vertices of $\mathbf{p}$. Since the tiling is locally finite, every neighborhood of the origin contains only finitely many vertices. Therefore there are two sequences of points of the curve, say $t_0, t_1, \ldots, t_k, \ldots$ and $w_0, w_1, \ldots, w_k, \ldots$, such that t_k and w_k tend to infinity in $\mathbb{R}^2$ and

$$(5.78) \qquad |f(t_k)| < \frac{1}{5}, \qquad |f(w_k)| > \frac{4}{5} \qquad \text{for every } k.$$

On the other hand, outside a neighborhood of the origin,

$$(5.79) \qquad\qquad |f(t) - f(w)| < (10\delta)^{-1}|t - w|.$$

Since the diameters of the tiles are bounded by δ, we see from (5.78) and (5.79) that there are two infinite sequences of vertices of $\mathbf{p}$, say $z_{n(k)}$ and $z_{m(k)}$, such that for all k

$$|\phi(z_{n(k)})| < \frac{2}{5} \quad , \qquad |\phi(z_{m(k)})| > \frac{3}{5}.$$

This proves the second step.

Now, it follows from Theorem (3.86) that $\mathbf{P}_o$ has extremal length ∞. By Corollary (3.48) Γ is recurrent. $\square$

REMARK. It follows that Grünbaum and Shephard's normal tilings are recurrent. It is not difficult to show that such tilings are uniformly embedded in $\mathbb{R}^2$. This contrasts with the situation of the hyperbolic disk: see §8 in Chapter IV.

CHAPTER VI

ROYDEN'S COMPACTIFICATION

§1. The algebra BD

The results on the discrete Dirichlet space **D** proved in Chapter III bear a close resemblance with results on the Dirichlet space of a Riemannian manifold. In this chapter we will stress this similarity by introducing the discrete analogue of the Royden and the harmonic boundaries of a Riemannian manifold (see Royden (1953), Costantinescu and Cornea (1963), Glasner and Katz (1970)). The definition and the study of Royden's boundary of a network are due to Kayano and Yamasaki. The first two sections are based on Yamasaki (1987), and the second half of the third section is based on the paper by Kayano and Yamasaki (1988). In the subsequent sections we will carry on the study of the Royden boundary of a network along the lines of the continuous theory, as described in Sario and Nakai (1970) and in Glasner and Katz (1970).

DEFINITION. Let (Γ, r) be a network. We denote by **BD** the space of all real, bounded and Dirichlet finite functions with the norm

$$\|u\| = \sup_{x \in V} |u(x)| + D(u)^{1/2}.$$

We set

$$\mathbf{BD}_0 = \mathbf{BD} \cap \mathbf{D}_0, \qquad \mathbf{HBD} = \mathbf{BD} \cap \mathbf{HD}.$$

Lemma (6.1). *Let $\{f_n\}$ be a sequence of functions in $\mathbf{D}_0$ which converges pointwise to a function f on V. If the sequence $\{D(f_n)\}$ is bounded, then $f \in \mathbf{D}_0$.*

Proof. Since the sequence $\{D(f_n)\}$ is bounded and $\mathbf{D}_0$ is a closed subspace, passing to a subsequence if necessary, f_n is weakly convergent to an element $\phi \in \mathbf{D}_0$. By Theorem (3.15), f_n converges pointwise to ϕ, whence the thesis. $\square$

Theorem (6.2). *The space $\mathbf{BD}$ is a commutative Banach algebra with respect to the pointwise product. The subspace $\mathbf{BD}_0$ is a closed ideal of $\mathbf{BD}$.*

Proof. Let $u, v \in \mathbf{BD}$. Set $\alpha = \sup_{x \in V} |u(x)|$ and $\beta = \sup_{x \in V} |v(x)|$. Then

$$\sum_{x \sim y} c(x,y)|u(x)v(x) - u(y)v(y)|^2 \leq \sum_{x \sim y} c(x,y)\big(\beta|u(x) - u(y)| + \alpha|v(y) - v(x)|\big)^2$$

$$\leq \big(\beta D(u)^{1/2} + \alpha D(v)^{1/2}\big)^2.$$

Thus

$$\|uv\| \leq \alpha\beta + \beta D(u)^{1/2} + \alpha D(v)^{1/2}$$

$$\leq (\alpha + D(u)^{1/2})(\beta + D(v)^{1/2}) = \|u\|\|v\|.$$

Therefore $\mathbf{BD}$ is a normed algebra.

Let $\{u_n\}$ be a Cauchy sequence in $\mathbf{BD}$. There exists $u \in \mathbf{D}$ such that $D(u - u_n) \to 0$ as $n \to \infty$. By Lemma (3.14), $u_n(x)$ converges to $u(x)$ for all $x \in V$. Since u_n is uniformly Cauchy, it follows that u is bounded. Hence $\mathbf{BD}$ is complete.

We now show that $\mathbf{BD}_0$ is an ideal. Assume that the network is transient (the recurrent case being obvious) and let $f \in \mathbf{BD}_0$ and $g \in \mathbf{BD}$. There is a sequence of finitely supported, uniformly bounded functions f_n such that $D(f_n)$ is bounded and $f_n \to f$ pointwise. To see this, set $M = \sup_{x \in V} |f(x)|$. Let $\{v_n\}$ be a sequence of finitely supported functions convergent to f in $\mathbf{D}$. In particular, $D(v_n)$ is bounded. Let $f_n = \max(\min(v_n, M), -M)$. Then f_n is finitely supported, $|f_n(x)| \leq M$ for all x and n and $D(f_n) \leq D(v_n)$ is bounded. Clearly, f_n converges to f pointwise.

The functions $f_n g$ are uniformly bounded, finitely supported and pointwise convergent to fg. Furthermore, we have as above

$$D(f_n g) \leq (M D(g)^{1/2} + \sup_x |g(x)| D(f_n)^{1/2})^2.$$

Hence the sequence $\{D(f_n g)\}$ is bounded and Lemma (6.1) implies that $fg \in \mathbf{BD}_0$. Finally, since the norm in $\mathbf{BD}$ is larger than the norm in $\mathbf{D}$, $\mathbf{BD}_0$ is closed in $\mathbf{BD}$. $\square$

DEFINITION. $\mathbf{BD}$ is called the Dirichlet algebra of the network (Γ, r).

We have already seen that Royden's decomposition theorem (Theorem (3.69)) is fundamental in the study of Dirichlet spaces. Here we have a refinement of Theorem (3.69).

Theorem (6.3). *Assume that (Γ, r) is transient and let F be a (possibly empty) subset of V. Then for every $f \in \mathbf{BD}$ there exist unique functions u and v with the following properties:*

(1) $f = u + v$;

(2) $v = 0$ on F and $v \in \mathbf{BD}_0(F)$, where $\mathbf{BD}_0(F)$ is the intersection of $\mathbf{BD}_0$ with the closure in $\mathbf{D}_0$ of the space of all real functions with finite support contained in $V \setminus F$;

(3) $u \in \mathbf{BD}$ and is harmonic on $V \setminus F$.

The function u satisfies the orthogonality relation

$$\lfloor u, \phi \rfloor = \frac{1}{2} \sum_{x \sim y} c(x,y)(u(x) - u(y))(\phi(x) - \phi(y)) = 0$$

for every $\phi \in \mathbf{BD}_0(F)$. Consequently, $D(f) = D(u) + D(v)$. Finally, if a and b are constants such that $a \geq f(x) \geq b$ on V, then $a \geq u(x) \geq b$ on V.

Proof. Let $\Gamma_n = (V_n, Y_n)$ be any exhaustion of Γ. For every n let f_n be the unique function which minimizes the strictly convex, positive functional D on the closed convex subset

$$\{g \in \mathbf{D} : g(x) = f(x) \text{ for } x \in V \setminus (V_n \setminus F)\}.$$

Then $f_n \leq a$. Otherwise, setting $f_n^* = \min(f_n, a)$, we would have $D(f_n^*) < D(f_n)$. Analogously, $b \leq f_n$.

Set $c_n = D(f_n)$. By the definition of f_n we have $c_m \leq c_n$ for $m \geq n$. Therefore c_n converges, say to c. Now,

$$c_m \leq D(2^{-1}(f_n + f_m)) \leq 4^{-1}(c_n + c_m + 2\sqrt{c_n c_m})$$

so that $D(2^{-1}(f_n + f_m)) \to c$ as $m, n \to \infty$. Moreover

$$c_m \leq D(2^{-1}(f_n + f_m)) \leq D(2^{-1}(f_n + f_m)) + D(2^{-1}(f_n - f_m))$$

$$= 2D(2^{-1}f_n) + 2D(2^{-1}f_m)$$

$$= (c_n + c_m)/2.$$

It follows that $D(f_n - f_m)$ tends to 0 as $m, n \to \infty$. Therefore f_n converges in $\mathbf{D}$ to a limit u. Clearly, $a \geq u(x) \geq b$ for all x, so that u is bounded. By the same argument used in the proof of Theorem (3.69) (f_n annihilates the Fréchet derivatives of D), we see that $\lfloor f_n, \phi \rfloor = 0$ for every $\phi \in \mathbf{BD}_0$ such that $\phi = 0$ on $V/(V_n/F)$. Letting n tend to infinity we see that $\lfloor u, \phi \rfloor = 0$ for e very $\phi \in \mathbf{BD}_0(F)$. In particular, taking $\phi = \delta_x$ we have $(1 - P)(u)(x) = 0$ for all $x \in V/F$. Hence u is harmonic on $V \setminus F$.

Now, put $v_n = f - f_n$ and $v = \lim_{n \to \infty} v_n$. Then $v \in \mathbf{BD}_0(F)$, since v_n is finitely supported and vanishes on F for all n.

Finally, let $h \in \mathbf{BD}$ be any function harmonic on $V \setminus F$ and let $g \in \mathbf{BD}_0(F)$. Then, as in Lemma (3.66), $\lfloor h, g \rfloor = 0$.

Hence, if $f = u' + v'$ is another decomposition of f satisfying (1), (2) and (3), then $u - u' = v - v'$ so that $u - u'$ vanishes on F and belongs to $\mathbf{BD}_0$. Hence $\lfloor u - u', u - u' \rfloor = 0$. It follows $u = u'$ and $v = v'$. $\square$

§2. Properties of Royden's compactification

Theorem (6.4). *Let (Γ, r) be an electrical network, where $\Gamma = (V, Y)$. There exists a unique (up to homeomorphisms) compact Hausdorff space $\Re$ with the following properties: $\Re$ contains V as an open dense subset, every function in $\mathbf{BD}$ can be continuously extended to $\Re$ and $\mathbf{BD}$ separates points in $\Re$.*

Proof. Uniqueness is obvious. Let us prove existence. Let $\Re$ denote the compact space (with the weak* topology) of all nontrivial multiplicative linear functionals on $\mathbf{BD}$ (see e.g. Yosida (1970), Larsen (1973)). For every $x \in V$ the map $f \mapsto f(x)$ is a linear multiplicative functional on $\mathbf{BD}$. Therefore $V \subset \Re$ and the Gelfand transform extends every $f \in \mathbf{BD}$ to a continuous function on $\Re$. Thus $\mathbf{BD}$ is a subalgebra of $C(\Re)$ (the space of all continuous real valued functions on $\Re$) and separates points in $\Re$. By the Stone–Weierstrass theorem $\mathbf{BD}$ is dense in $C(\Re)$. It follows that V is dense in $\Re$. Finally, the function δ_x is continuous on $\Re$ for every $x \in V$. Hence x is isolated in $\Re$ and V is open. $\square$

Corollary (6.5). $\mathbf{BD}$ *is dense in the space $C(\Re)$ of all continuous real–valued functions on $\Re$.*

In the following we will still denote by f the extension of $f \in \mathbf{BD}$ to the whole of $\Re$.

DEFINITION. The compact space $\Re$ in Theorem (6.4) is called the Royden compactification of the network (Γ, r). The compact subset $b\Re = \Re \setminus V$ is called the Royden boundary of (Γ, r). We distinguish the following important part of Royden's boundary

$$\Delta = \{x \in b\Re : f(x) = 0 \quad \text{for all } f \in \mathbf{BD}_0\}.$$

The set Δ is called the harmonic boundary of the network (Γ, r). Note that $\Delta = \emptyset$ if (Γ, r) is recurrent. Conversely, if $\Delta = \emptyset$, then (Γ, r) is recurrent by Lemma (6.6) below (take $F = \Re$) and by Theorem (3.63).

Lemma (6.6). *For any closed subset F of $\Re$ such that $F \cap \Delta = \emptyset$, there exists $f \in \mathbf{BD}_0$ such that $f = 1$ on F and $0 \le f \le 1$ on $\Re$.*

For any two nonempty disjoint compact subsets F_1 and F_2 of $\Re$ and any two real numbers a_1 and a_2, there exists $f \in \mathbf{BD}$ such that $f = a_j$ on F_j and $\min(a_1, a_2) \le f \le \max(a_1, a_2)$.

Proof. Since $F \cap \Delta = \emptyset$, for every $x \in F$ there exists $f_x \in \mathbf{BD}_0$ such that $f_x(x) \neq 0$. Since $\mathbf{BD}_0$ is an ideal, we may assume $f_x \geq 0$ on V and $f_x(x) > 0$. Let U_x be a neighborhood of x in $\mathfrak{R}$ such that $f_x(y) > 0$ for $y \in U_x$. By compactness there exist $x_1, \ldots, x_n$ such that $F \subseteq \cup_{j=1}^n U_{x_j}$. Put

$$g = \sum_{j=1}^n f_{x_j}, \quad \alpha = \inf\{g(x) : x \in F\}.$$

Then $g \in \mathbf{BD}_0$ and $\alpha > 0$. The function $f = \min(1, \alpha^{-1}g)$ is the function looked for, provided that we show that f belongs to $\mathbf{BD}_0$. Let g_n be finitely supported functions such that $D(g - g_n) \to 0$ as $n \to \infty$. Let $f_n = \min(1, \alpha^{-1}g_n)$. Then $D(f_n)$ is bounded and f_n converges pointwise to f on V. By Lemma (6.1) $f \in \mathbf{BD}_0$.

To prove the second assertion we may assume $a_1 > a_2$. By Urisohn's lemma there exists $\phi \in C(\mathfrak{R})$ such that $\phi = a_1 + 2$ on F_1 and $\phi = a_2 - 2$ on F_2. Since $\mathbf{BD}$ is dense in $C(\mathfrak{R})$, there is $g \in \mathbf{BD}$ such that $|\phi - g| < 1$. The function $f = \max(\min(a_1, g), a_2)$ belongs to $\mathbf{BD}$ and has the required property. $\square$

We have the following minimum principle in $\mathbf{HBD}$.

Theorem (6.7). *Let (Γ, r) be transient. If $u \in \mathbf{HBD}$ is not identically zero and $u \geq 0$ on Δ, then $u > 0$ on V.*

Proof. For every $\epsilon > 0$ set $F_\epsilon = \{x \in \mathfrak{R} : u(x) + \epsilon \leq 0\}$. Then F_ϵ is closed and $F_\epsilon \cap \Delta = \emptyset$. By Lemma (6.6) there exists $f \in \mathbf{BD}_0$ such that $f = 1$ on F_ϵ and $0 \leq f \leq 1$ on $\mathfrak{R}$.

Since u is bounded, $u \geq -c$ for some constant $c > 0$. Therefore $u + \epsilon + cf \geq 0$ on $\mathfrak{R}$. By Theorem (6.3) (with $F = \emptyset$) $u + \epsilon \geq 0$ on V. By the arbitrariety of ϵ, $u \geq 0$ on V. By the minimum principle for harmonic functions, $u > 0$ on V. $\square$

As a consequence of Theorem (6.7) we have that only a function in $\mathbf{BD}_0$ can vanish on Δ.

Corollary (6.8). *Let (Γ, r) be a transient network. Then*

$$\mathbf{BD}_0 = \{f \in \mathbf{BD} : f(x) = 0 \quad \text{for all } x \in \Delta\}.$$

Proof. Put $\mathbf{X} = \{f \in \mathbf{BD} : f(x) = 0 \quad \text{for all } x \in \Delta\}$. Then $\mathbf{BD}_0 \subseteq \mathbf{X}$. Let $f \in \mathbf{X}$. There are $u \in \mathbf{HBD}$ and $v \in \mathbf{BD}_0$ such that $f = u + v$. Since $v = 0$ on Δ, Theorem (6.7) implies that $u = 0$ on $\mathfrak{R}$. Hence $f \in \mathbf{BD}_0$. $\square$

It turns out that every Dirichlet finite, not necessarily bounded, function can be extended to $\mathfrak{R}$.

Theorem (6.9). *Every $f \in \mathbf{D}$ has a continuous extension to a function defined on $\mathfrak{R}$ with values in $[-\infty, +\infty]$.*

Proof. Suppose first $f \geq 0$. Set $f_n = \min(n, f)$. Then $f_n \in \mathbf{BD}$ and $f_{n+1} \geq f_n$ on V. By density $f_{n+1} \geq f_n$ on $\mathfrak{R}$. Hence $f(x) = \lim_{n \to \infty} f_n(x)$ exists for all $x \in \mathfrak{R}$. We have $\min((f, m), n) = \min(f, n)$ for $m > n$ on V, whence we get

(6.10) $\min(f_m, n) = f_n$ for $m > n$ on $\mathfrak{R}$.

Suppose now $z \in \mathfrak{R}$ and $f(z) < \infty$. Let $n > f(z)$. Then $f_n(z) < n$ and there is a neighborhood U of z such that $f_n(x) < n$ on U. By (6.10) $\min(f_m(x), n) = f_n(x) < n$ all $m > n$ on U. Hence $f_m(x) = f_n(x)$ for $m > n$ on U and f is continuous on U.

Next assume that $f(z) = +\infty$. For every $c > 0$ there is n such that $f_n(z) > c$. As before there is a neighborhood U of z such that $f_n(x) > c$ for on U. Then $f_m(x) > c$ for $m > n$ on U. Hence $f(x) > c$ for all $x \in U$ i.e., f is continuous at z.

In the general case write f as sum of its positive and negative part $f^+ = \max(f, 0)$ and $f^- = -\min(f, 0)$. It is easily seen that $f^+ - f^-$ has a definite meaning in $\mathfrak{R}$. In fact if, say, $f^+(z) = +\infty$, then there is neighborhood U of z such that $f^-(x) = 0$ for all $x \in U \cap V$. Hence $f^-(z) = 0$. $\square$

The minimum principle for functions in **HBD** holds true also for functions in **HD**.

Corollary (6.11). *Let $f \in \mathbf{HD}$ be not identically zero and such that $f \geq 0$ on Δ. Then $f > 0$ on V.*

Proof. Let f_n be as in the proof of Theorem (6.9). The argument used to establish the continuity of f on $\mathfrak{R}$ shows that $f_n \geq 0$ on Δ. Set $f_n = u_n + v_n$ with $u_n \in \mathbf{HBD}$ and $v_n \in \mathbf{BD}_0$. As $D(u_n) \leq D(f_n) \leq D(f)$, we see that, passing to a subsequence if necessary, $\{u_n\}$ is weakly convergent to some function $\phi \in \mathbf{HD}$ (and v_n is weakly convergent to an element of $\mathbf{D}_0$). By Theorem (3.15) we see that $\phi = f$.

Since $u_n = f_n$ on Δ, by applying Theorem (6.7) we see that $u_n > 0$ on V. It follows $f > 0$ on V. $\square$

We have the following extension to $\mathbf{D}$ of Theorem (6.8).

Theorem (6.12). *Suppose that (Γ, r) is transient. Then*

$$\mathbf{D}_0 = \{f \in \mathbf{D} : f(x) = 0 \quad \text{for all } x \in \Delta\}.$$

Proof. Assume $f \in \mathbf{D}_0$. By considering the positive and negative part of f (which belong to $\mathbf{D}_0$ by Lemma (3.70)), we may restrict our attention to the case where $f \geq 0$. For every positive integer n we set, as in the proof of the above Theorem (6.9), $f_n = \min(f, n)$. Then $f_n \in \mathbf{BD}_0$ (as shown in the proof of Lemma (6.6)). So, $f_n = 0$ on Δ, whence $f = 0$ on Δ.

Conversely, assume $f \geq 0$ and $f = 0$ on Δ. Then $f_n = 0$ on Δ for all n. By Corollary (6.8) $f_n \in \mathbf{BD}_0$. By Lemma (6.1) $f \in \mathbf{D}_0$. $\square$

The following Lemma proves the existence of the discrete analogue of Evan's superharmonic function (see Constantinescu (1962) for Riemann surfaces).

Lemma (6.13). *Suppose that (Γ, r) is transient and let F be a nonempty compact subset of $b\mathfrak{R} \setminus \Delta$. Then there exists a nonnegative superharmonic function $g \in \mathbf{D}_0$ such that $g = +\infty$ on F.*

Proof. By Lemma (6.6) there exists a compact set $K \subset \mathfrak{R}$ such that F is contained in the interior of K and such that $K \cap \Delta = \emptyset$. By Lemma (6.6) again there exists $f \in \mathbf{BD}_0$ such that $f = 1$ on K and $0 \leq f \leq 1$.

Let $\Gamma_n = (V_n, Y_n)$ be an exhaustion of Γ. Set $K_n = K \setminus V_n$ and apply Theorem (6.3). There exist functions u_n and v_n such that $f = u_n + v_n$ for every n, u_n is harmonic on $V \setminus K_n$, v_n belongs to $\mathbf{BD}_0$ and $v_n = 0$ on K_n. Then $u_n = f - v_n$ belongs to $\mathbf{BD}_0$ and $0 \leq u_n(x) \leq 1$ on V. Since $u_n(x) = f(x) = 1$ on K_n, we see that u_n is a positive superharmonic function on V.

Always by Theorem (6.3) we have

$$(6.14) \qquad D(f) = D(u_n) + D(v_n) \quad \text{for every } n,$$

so that $D(u_n)$ is bounded. Passing to a subsequence if necessary, u_n is weakly convergent (in $\mathbf{D}$) to some $u \in \mathbf{D}_0$. Since $0 \leq u_n(x) \leq 1$, we can apply the dominated convergence theorem for every $x \in V$

$$\sum_{y \sim x} p(x, y) u(y) = \lim_{n \to \infty} \sum_{y \sim x} p(x, y) u_n(y) = \lim_{n \to \infty} u_n(x) = u(x).$$

Hence u is harmonic on the whole of V. As $u \in \mathbf{D}_0$, we have $u = 0$. It follows $\lim_{n \to \infty} v_n(x) = f(x)$ for all x. We have by Fatou's lemma

$$D(f) \leq \liminf_{n \to \infty} D(v_n).$$

Taking (6.14) into account we may conclude that $D(v_n) \to D(f)$ and $D(u_n) \to 0$ as $n \to \infty$. Passing to a further subsequence if necessary, we may assume that $\sum_{n=1}^{\infty} u_n$ is absolutely convergent in the norm of $\mathbf{D}$.

Let $g_m = \sum_{n=1}^{m} u_n$ and $g = \sum_{n=1}^{\infty} u_n$. Then $g \in \mathbf{D}_0$. Since g_m is positive and superharmonic, by Fatou's lemma g is positive and superharmonic. Furthermore $g_m = m$ on K_m, so that $g \geq g_m \geq m$ for all m on F. Hence $g = +\infty$ on F. $\square$

Now we shall study the "points at infinity" of a family of paths. For a set $A \in \mathfrak{R}$ let $\overline{A}$ denote the closure of A in $\mathfrak{R}$.

DEFINITION. Let (Γ, r) be a network and let $\mathbf{p}$ be a one-sided infinite path in Γ. Let $V(\mathbf{p})$ denote the vertex set of the path and $\overline{V}(\mathbf{p})$ the closure in $\mathfrak{R}$ of $V(\mathbf{p})$. The set of the extreme points $E(\mathbf{p})$ of $\mathbf{p}$ is defined as

$$(6.15) \qquad E(\mathbf{p}) = \overline{V}(\mathbf{p}) \cap b\mathfrak{R}.$$

Theorem (6.16). *Let (Γ, r) be a transient network and let $\mathbf{P}$ be a family of one–sided infinite paths in Γ such that*

$$F = \overline{\bigcup_{\mathbf{p} \in \mathbf{P}} E(\mathbf{p})}$$

is disjoint from Δ. Then $\lambda(\mathbf{P}) = \infty$ (where λ denotes the extremal length).

Proof. By the preceding Lemma (6.13) we can find $g \in \mathbf{D}_0$ such that $g = +\infty$ on F. By Theorem (6.12) $g = 0$ on Δ. Let $x_0 \sim x_1 \sim \cdots$ be the vertices of a path $\mathbf{p} \in \mathbf{P}$. Since $E(\mathbf{p}) \subseteq F$ we have that $g(\mathbf{p}) = \lim_{j \to \infty} g(x_j) = +\infty$. Thus

$$(6.17) \quad \sum_{j=1}^{\infty} r(x_{j-1}, x_j)|c(x_{j-1}, x_j)(g(x_{j-1}) - g(x_j))| \geq \lim_{n \to \infty}(g(x_n) - g(x_0)) = +\infty.$$

By (3) of Lemma (3.75) we get $\lambda(\mathbf{P}) = \infty$. $\square$

As a converse of this result, we can prove the following theorem.

Theorem (6.18). *Let (Γ, r) be a locally finite transient network. Let $\mathbf{P}$ be the family of all one–sided infinite paths in Γ and let $\mathbf{P}_\infty \subset \mathbf{P}$ be any subfamily such that $\lambda(\mathbf{P}_\infty) = \infty$. Then*

$$(6.19) \qquad \Delta \subseteq \overline{\{\cup_{\mathbf{p}} E(\mathbf{p}) : \mathbf{p} \in \mathbf{P}/\mathbf{P}_\infty\}}.$$

Proof. Set $F = \overline{\{\cup_{\mathbf{p}} E(\mathbf{p}) : \mathbf{p} \in \mathbf{P}/\mathbf{P}_\infty\}}$. By Corollary (3.84), $\lambda(\mathbf{P}/\mathbf{P}_\infty) < \infty$. It follows from Theorem (6.16) above that $F \cap \Delta \neq \emptyset$.

Assume, by way of contradiction, that there exists a point $z \in \Delta$ which is not in F. We claim that there exists $u \in \mathbf{HD}$ such that $0 < u < 1$ on V, $u(z) = 1$ and $u = 0$ on $\Delta \cap F$. In fact, by Lemma (6.6) we can find $g \in \mathbf{BD}$ such that $0 \leq g \leq 1$ on V, $g(z) = 1$ and $g = 0$ on $\Delta \cap F$. We decompose g as $v + u$, with $v \in \mathbf{BD}_0$ and $u \in \mathbf{HBD}$. The function u meets our requirements by Theorem (6.7).

In the proof of Theorem (3.93) we showed that there exists a subnetwork (Γ^+, r^+) (r^+ being the restriction of r to the edge set of the subgraph Γ^+) and a vertex a of Γ^+ such that $\lambda(\mathbf{P}_a^+)$ is finite and $u(\mathbf{p}) > u(a)$ for all $\mathbf{p} \in \mathbf{P}_a^+$ (here $\mathbf{P}_a^+$ is the set of all one–sided paths in Γ^+ starting at a).

Let us put, in analogy with (6.19),

$$F_1 = \overline{\{\bigcup E(\mathbf{p}) : \mathbf{p} \in \mathbf{P}_a^+/\mathbf{P}_\infty\}}.$$

Since $\lambda(\mathbf{P}_a^+/\mathbf{P}_\infty) < \infty$, we get $\Delta \cap F_1 \neq \emptyset$, by Theorem (6.16). Since $F_1 \subseteq F$, u vanishes on $F_1 \cap \Delta$. But $u(\mathbf{p}) > u(a) > 0$ for $\mathbf{p} \in \mathbf{P}_a^+$, which in turn implies $u(x) \geq u(a)$ for all $x \in F_1$, a contradiction. Thus F contains the harmonic boundary Δ. $\square$

§3. Boundary points

We begin this section by studying the isolated points in the harmonic boundary. DEFINITION. Let (Γ, r) be a transient network. A nonnegative function $u \in \mathbf{HD}$ is called $\mathbf{HD}$–minimal if u is not identically zero and, for every $g \in \mathbf{HD}$ such that $u \geq g \geq 0$ on V, there exists a constants c_g such that $g = c_g u$.

$\mathbf{HD}$–minimality is characterized, as in the continuous case, in terms of the harmonic boundary. The following theorem is proved exactly as in the continuous case.

Theorem (6.20). *Let (Γ, r) be a transient network. A function $u \in \mathbf{HD}$ is minimal if and only if there exists a point $z \in \Delta$, which is isolated for the relative topology of Δ, such that*

$$(6.21) \qquad u(z) > 0, \quad u(x) = 0 \text{ for all } x \in \Delta \setminus \{z\}.$$

Moreover, a $\mathbf{HD}$–minimal function is bounded and strictly positive on V.

Proof. Let z be an isolated point of Δ. Suppose that $u \in \mathbf{HD}$ is such that $u(z) > 0$ and $u(x) = 0$ at the points of the harmonic boundary other than z. By Lemma (6.6) there exists a nonnegative $g \in \mathbf{BD}$ such that $g(z) = 1$ and $g = 0$ on $\Delta \setminus \{z\}$. By Theorem (6.3) (with $F = \emptyset$) and the definition of Δ, we can find a function $f \in \mathbf{HBD}$ such that $f(z) = 1$ and $f = 0$ on $\Delta \setminus \{z\}$. By the minimum principle (Theorem (6.7)), $f > 0$ on V.

For every c, $0 < c < u(z)$, $u - cf$ is nonnegative on the harmonic boundary and belongs to $\mathbf{HD}$. Thus, by the minimum principle in $\mathbf{HD}$ (Corollary (6.11)), $u > cf$ on V. Fix any point x_0 in V. We have $c < u(x_0)/f(x_0)$. This implies that c can not be chosen arbitrarily large i.e., $0 < u(z) < \infty$. Hence, by Corollary (6.11) again, u is bounded and positive on V.

For any $g \in \mathbf{HD}$ with $u \geq g \geq 0$ on V we have $g = 0$ on $\Delta \setminus \{z\}$ and $g(z) < \infty$. As a consequence $c_g u - g$, with $c_g = g(z)/u(z)$, vanishes on the harmonic boundary. Therefore $c_g u - g = 0$ everywhere.

Conversely suppose that u is $\mathbf{HD}$–minimal. For some $z \in \Delta$ we have $u(z) > 0$. Assume, by contradiction, that there exists another point $t \in \Delta$ such that $u(t) > 0$. Then, there exists $f \in \mathbf{BD}$, $0 \leq f \leq 1$, such that $f(t) = 1$ and $f(z) = 0$. Let $fu = u' + v$ be the Royden decomposition of fu, with $u' \in \mathbf{HBD}$ not identically 0. Then $0 \leq u' = fu \leq u$ on the harmonic boundary. Therefore $0 \leq u' \leq u$ on $\Re$ and there is a constant $c > 0$ such that $u' = cu$. This is absurd since $u'(z) = 0$ while $u(z) > 0$. Hence z is isolated by the continuity of u. By the first part of the proof we have that $u \in \mathbf{HBD}$. □

Corollary (6.22). *Let z be any isolated point of the harmonic boundary. There always exists a function $u \in \mathbf{HBD}$ such that $u(z) = 1$ and $u = 0$ on Δ. Moreover every function g in $\mathbf{HD}$ has finite value at z.*

Proof. Only the last sentence needs a proof. Let f be the harmonic component of g^2 in the Royden decomposition of such a function. Suppose that $f(z) = +\infty$. Let u be the $\mathbf{HD}$–minimal function such that $u(z) = 1$. Then $f - nu \geq 0$ on Δ for all $n > 0$. Hence $f(x) > nu(x)$ for all $x \in V$ and $n > 0$. It follows $f(x) = +\infty$, which is absurd. □

REMARK. Note that the set of isolated points of the harmonic boundary may be empty; see Example (6.37) below.

It is known in the case of Riemannian surfaces that the singleton of a point in $b\Re$ is not a G_δ set. In the discrete case the situation is quite different. On one hand it is still true that singletons of elements in $b\Re \setminus \Delta$ are not G_δ sets. However, in contrast with the continuous case, it is easy to produce examples of networks such that $\{x\}$ is a G_δ set for some $x \in \Delta$ (see Example (6.36) below).

Another behavior with no continuous counterpart is exhibited in Example (6.37) below.

We start with two technical lemmas.

Lemma (6.23). *Assume that (Γ, r) is a locally finite network. Let $u \in \mathbf{BD}_0$ and let A be a finite subset of V. For every $\epsilon > 0$ there exists a finitely supported f such that $f = u$ on A, $\sup_{x \in V} |f(x)| \leq \sup_{x \in V} |u(x)|$ and $D(u - f) < \epsilon$.*

Proof. Let $\{f_n\}$ be a sequence of finitely supported functions such that $f_n \to f$ in $\mathbf{D}$. The functions f_n can also be chosen in such a way that $|f_n(x)| \leq \sup_{x \in V} |u(x)|$

for all x and n. Set $u_n = u$ on A and $u_n = f_n$ on $V \setminus A$. We have only to show that $D(u_n - u) \to 0$ as $n \to \infty$.

Since $f_n(x) \to u(x)$ for every $x \in V$, $(u - u_n)(x) - (u - u_n)(y) \to 0$ for every couple of neighboring vertices x and y. We have

$$(6.24) \quad D(u - u_n) \le \sum_{\substack{x \in A, y \notin A \\ x \sim y}} c(x,y)((u - u_n)(x) - (u - u_n)(y))^2 + D(u - f_n).$$

By the finiteness of A and the definition of f_n the terms on the right hand side of (6.24) tend to 0. $\square$

Lemma (6.25). *Assume that (Γ, r) is a locally finite network. Let $\{V_n\}$ be a sequence of infinite subsets of V. For every $u \in \mathbf{BD}_0$ there exist $\phi \in \mathbf{BD}_0$ and two sequences of nodes a_n and b_n satisfying the following conditions*

$$
\begin{aligned}
&a_m \in V_{2m-1}, \quad b_m \in V_{2m}, \quad m = 1, 2, \dots \\
(6.26) \quad &a_n \ne a_m, \quad b_n \ne b_m \quad \text{for } n \ne m, \\
&\phi(a_m) = 0 \quad \text{and} \quad \phi(b_m) = u(b_m) \quad \text{for all } m.
\end{aligned}
$$

Proof. Choose any $x_1 \in V_1$ and set $A_1 = \{x_1\}$. By Lemma (6.23) there exists a finitely supported f_1 such that $f_1(x_1) = u(x_1)$, $\sup_x |f_1(x)| \le \sup_x |u(x)|$ and $D(u - f_1) < 1/2$.

For $x \in V$ and $c > 0$ let $B(x, c)$ denote the ball of radius c and center x. For every subset $U \subseteq V$ we set

$$C(U) = \bigcup_{x \in U} B(x, 1).$$

For every function f on V denote by $s(f)$ the support of f. We set $B_1 = C(s(f_1)) \cup A_1$. Choose $x_2 \in V_2 \setminus B_1$ and set $A_2 = C(B_1) \cup \{x_2\}$. By Lemma (6.23) again, there exists a finitely supported f_2 such that $f_2 = u$ on A_2, $\sup_x |f_2(x)| \le \sup_x |u(x)|$ and $D(u - f_2) < 1/4$. Repeating this construction we end up with a sequence of points x_n, two sequences of finite sets A_n and B_n and a sequence of finitely supported functions f_n such that

$$
\begin{aligned}
(6.27) \quad &x_n \in V_n \setminus B_{n-1}, \quad \text{where } B_n = C(s(f_n)) \cup A_n, \quad A_n = C(B_{n-1}) \cup \{x_n\}, \\
&f_n = u \text{ on } A_n, \quad \sup |f_n| \le \sup |u| \text{ and } D(u - f_n) < 2^{-n}.
\end{aligned}
$$

Set $B_0 = \emptyset$. For $n \ge 1$ every B_n is finite, since the network is locally finite. Note that $C(B_{n-1}) \subseteq B_n$, $x_n \in B_n \setminus B_{n-1}$ and $f(x_n) = u(x_n)$, $n = 1, 2, \dots$.

We define $a_m = x_{2m-1}$ and $b_m = x_{2m}$. By (6.27) a_m and b_m satisfy the first and the second conditions of (6.26). Let us construct the function ϕ. For every $k \geq 1$ we set

$$\phi_k(x) = u(x) - f_{2m-1}(x) \quad \text{if } x \in B_{2m-1} \setminus B_{2m-2}, \text{ for } m = 1, 2, \ldots, k$$
$$(6.28) \quad \phi_k(x) = f_{2m}(x) \quad \text{if } x \in B_{2m} \setminus B_{2m-1}, \quad \text{for } m = 1, 2, \ldots, k$$
$$\phi_k(x) = 0 \quad \text{if } x \in V \setminus B_{2m}.$$

It is clear that ϕ_k is pointwise convergent to a function ϕ satisfying the third condition of (6.26). In order to show that ϕ is in $\mathbf{BD}_0$ we evaluate $D(\phi_k)$. Set, for all $n = 1, 2, \ldots,$

$$Y_n = \{[x, y] \in Y : x, y \in B_n \setminus B_{n-1}\}$$
$$Y_n' = \{[x, y] \in Y : \text{either } x \in B_n, y \notin B_n, \text{ or } y \in B_n, x \notin B_n\}.$$

Since $C(B_n) \subset B_{n+1}$, we see that $\bigcup_{n=1}^{2k}(Y_n \cup Y_n')$ is exactly the set of all edges having at least one endpoint in B_{2k}. It follows

$$
2D(\phi_k) = \sum_{n=1}^{k} \sum_{[x,y] \in Y_n} c(x,y)((\phi_k(x) - \phi_k(y))^2
$$
$$
(6.29)
$$
$$
+ \sum_{n=1}^{k} \sum_{[x,y] \in Y_n'} c(x,y)(\phi_k(x) - \phi_k(y))^2.
$$

By (6.28) we have, if $[x, y] \in Y_{2m-1}$ and $k \geq m$,

$$(6.30) \qquad (\phi_k(x) - \phi_k(y))^2 = ((f_{2m-1} - u)(x) - (f_{2m-1} - u)(y))^2.$$

On the other hand, if $[x, y] \in Y_{2m}$ and $m \leq k$, then we get

$$
(6.31) \quad
\begin{aligned}
(\phi_k(x) - \phi_k(y))^2 &= (f_{2m}(x) - f_{2m}(y))^2 \\
&\leq 2((f_{2m} - u)(x) - (f_{2m} - u)(y))^2 + 2(u(x) - u(y))^2.
\end{aligned}
$$

If $[x, y] \in Y_n'$ with $n \leq 2k$, $x \in B_n, y \notin B_n$, then $x \notin s(f_n)$ for, otherwise, $y \in C(s(f_n)) \subset B_n$. Hence $f_n(x) = 0$ and $f_{n+1}(y) = u(y)$. It follows that $\phi_k(x) = \phi_k(y) = 0$ if k is even, $\phi_k(x) = u(x)$ and $\phi_k(y) = u(y)$ if k is odd. Collecting (6.28), (6.29), (6.30) and (6.31) we have

$$
\begin{aligned}
2D(\phi_k) &\leq 2D(u) + 2\sum_{n=1}^{2k} D(u - f_n) \leq 2D(u) + 2\sum_{n=1}^{2k} 2^{-n} \\
&\leq 2D(u) + 2
\end{aligned}
$$

for all k. Since $\sup |\phi_k| \leq 2 \sup |u|$, $\phi \in \mathbf{BD}_0$ by Lemma (6.1). $\square$

Theorem (6.32). *Let (Γ, r) be an infinite locally finite network. For any $z \in \mathfrak{R} \setminus \Delta$ the singleton $\{z\}$ is not a G_δ set.*

Proof. Assume, by contradiction, that $\{z\}$ is a G_δ set for some $z \in b\mathfrak{R} \setminus \Delta$. Then there is a sequence of open neighborhoods U_n of z in $\mathfrak{R}$ such that $\overline{U}_{n+1} \subset U_n$ and $\cap_n U_n = \{z\}$. By Lemma (6.6) we can find a function $u \in \mathbf{BD}_0$ such that $u(z) = 1$ and $0 \le u \le 1$. Set

$$V_n = \{x \in U_n \cap V : u(x) > 1/2\}, \quad \text{for } n = 1, 2, \ldots.$$

Each V_n is infinite and we may apply Lemma (6.25). Let a_n, b_n and $\phi \in \mathbf{BD}_0$ be as in (6.26). Then $\phi(a_m) = 0$ and $\phi(b_m) > 1/2$. On the other hand, both sequences a_n and b_n converge to z in $\mathfrak{R}$. This contradicts the continuity of ϕ and establishes the thesis. $\square$

Next, we prove a criterion for a singleton in the harmonic boundary of a not necessarily locally finite network to be a G_δ set. Note that such a criterion has no continuous counterpart.

Theorem (6.33). *Let $z \in \Delta$. The singleton $\{z\}$ is a G_δ set if and only if there exists $f \in \mathbf{BD}$ such that $f(z) = 1$ and $0 \le f(x) < 1$ for all $x \in \mathfrak{R}$, $x \ne z$.*

Proof. The "if" part is clear since f is continuous on $\mathfrak{R}$. Conversely, suppose that $\{U_n\}$ is a sequence of open sets in $\mathfrak{R}$ such that $\cap_{n=1}^{\infty} U_n = \{z\}$. Set $F_n = \mathfrak{R} \setminus U_n$. The set F_n is compact and does not contain z. By Lemma (6.6) there exists $f_n \in \mathbf{BD}$ such that $f_n(z) = 1$, $f_n(x) = 0$ for $x \in F_n$ and $0 \le f_n \le 1$. Since $\cup_{n=1}^{\infty} F_n = \mathfrak{R} \setminus \{z\}$, the function $f = \sum_{n=1}^{\infty} 2^{-n} f_n$ has the required properties. $\square$

Theorem (6.34). *Let (Γ, r) be an infinite network and let $\mathbf{p}$ be a one–sided infinite path with vertices $x_0 \sim x_1 \sim \cdots$. If $\sum_{j=1}^{\infty} r(x_{j-1}, x_j) < \infty$, then $E(\mathbf{p})$ is a singleton and $E(\mathbf{p}) \subset \Delta$.*

Proof. Let f be any function in $\mathbf{BD}$. Then for all $n > m$,

$$|f(x_n) - f(x_m)| \le \sum_{j=m+1}^{n} |f(x_{j-1}) - f(x_j)|$$

$$\le \Big(\sum_{j=m+1}^{n} r(x_{j-1}, x_j)\Big)^{1/2} \Big(\sum_{j=m+1}^{n} c(x_{j-1}, x_j)|f(x_{j-1}) - f(x_j)|^2\Big)^{1/2}.$$

It follows that $f(x_n)$ converges to a limit $f(\mathbf{p})$. Since functions in $\mathbf{BD}$ separate the points in $\mathfrak{R}$, we conclude that $E(\mathbf{p})$ is a singleton.

Next, we claim that the extremal length of the set $\{\mathbf{p}\}$ containing the single path $\mathbf{p}$ is finite. Let I be any chain with finite energy. We have

$$\sum_{j=1}^{\infty} r(x_{j-1}, x_j) |i(x_{j-1}, x_j)| \leq (\mathcal{W}(I))^{1/2} \left(\sum_{j=1}^{\infty} r(x_{j-1}, x_j) \right)^{1/2}$$
$$< \infty.$$

By (3) of Lemma (3.75) we have $\lambda(\{\mathbf{p}\}) < \infty$. By Theorem (3.86), $f(\mathbf{p}) = 0$ for every $f \in \mathbf{BD}_0$. Finally, by Corollary (6.8), $E(\mathbf{p}) \subset \Delta$. $\square$

EXAMPLE (6.35). Let Γ be a one-sided path $\mathbf{p}$ with vertices $x_0 \sim x_1 \sim \cdots$. Let the resistances on Γ be assigned in such a way that $\sum_{j=1}^{\infty} r(x_{j-1}, x_j) < \infty$, as in the Theorem (6.34). Then $b\mathfrak{R}$ reduces to the singleton $E(\mathbf{p}) = \Delta$.

We claim that $E(\mathbf{p})$ is a G_δ set. Let g be the function on V such that $g(x_0) = 0$ and $g(x_n) = \sum_{j=1}^{n} r(x_{j-1}, x_j)$, $n = 1, 2, \ldots$. Then $D(g) = \sum_{j=1}^{\infty} r(x_{j-1}, x_j) < \infty$, so that $g \in \mathbf{BD}$. Note that $g(\mathbf{p}) = D(g)$. Put $f = g/D(g)$. By Theorem (6.33) $E(\mathbf{p})$ is a G_δ set.

EXAMPLE (6.36). Let Γ be as in Example (6.35). This time we assign resistance 1 to every edge so as to obtain the recurrent network $\mathbb{Z}_+$. We show that $b\mathfrak{R}$ is uncountable.

Define a function f on the vertices x_k in the following way

$$f(x_k) = t2^{-k} \quad \text{for } k = 2^{m+1} + t - 2 \text{ and } t = 0, 1, \ldots, 2^m,$$
$$f(x_k) = 1 - t2^{-m} \quad \text{for } k = 2^{m+1} + 2^m + t - 2 \text{ and } t = 1, \ldots, 2^m,$$

where $m = 0, 1, \ldots$. Then we have

$$D(f) = \sum_{j=0}^{\infty} |f(x_j) - f(x_{j-1})|^2 = 2 \sum_{m=0}^{\infty} 2^{-2m} \sum_{j=1}^{2^m} 1 < \infty.$$

Hence $f \in \mathbf{BD}$. The set $f(V)$ is dense in the interval $[0,1]$. Therefore $f(\mathfrak{R}) = [0,1]$, by the continuity of f on $\mathfrak{R}$. As $f(V)$ is countable, $b\mathfrak{R}$ is uncountable.

EXAMPLE (6.37). Let now Γ be a locally finite graph of bounded degree satisfying a strong isoperimetric inequality. We assign resistance 1 to every edge.

By (4.28) $f(x) \to 0$ as $x \to \infty$ for every $f \in \mathbf{BD}_0$. Thus $b\mathfrak{R} = \Delta$ by Corollary (6.8). Note that this example has no analogue in the continuous case. In fact, in the case of a hyperbolic Riemann surface $b\mathfrak{R} \setminus \Delta$ is dense in $b\mathfrak{R}$ (see Sario and Nakai (1970), page 157).

By Theorem (6.18) we have immediately

(6.38) $$b\mathfrak{R} = \Delta = \overline{\{\bigcup E(\mathbf{p}) : \mathbf{p} \in \mathbf{P}\}},$$

where $\mathbf{P}$ is the set of all one–sided infinite paths in Γ.

In the case of a homogeneous tree T we can say more. Suppose that Γ is a homogeneous tree T of degree $q \geq 3$. Let us fix a reference vertex o. For every vertex $x \neq o$ there is a unique infinite subtree induced by x and by all the vertices y such that $x \succ y$. Remember that $x \succ y$ if x is on the unique path joining o and y (see §2 in Chapter II). We shall denote by T_x this subtree. The graph induced by the vertices not in T_x is another infinite subtree which will be denoted by $T \setminus T_x$.

By Theorem (5.72) T satisfies a strong isoperimetric inequality, so that (6.38) holds. Suppose that $\mathbf{p}$ and $\mathbf{p}'$ are distinct one–sided infinite paths starting from o. Then there is a vertex x such that, with the possible exception of finitely many nodes, $V(\mathbf{p}) \subset T_x$ and $V(\mathbf{p}') \subset T \setminus T_x$, where $V(\mathbf{p})$ is the vertex set of $\mathbf{p}$ and $V(\mathbf{p}')$ is the vertex set of $\mathbf{p}'$. Define a function u by setting $u(x) = 1$ on the vertices of T_x and $u(x) = 0$ on the vertices of $T \setminus T_x$. Then $u \in \mathbf{BD}$ and u separates $E(\mathbf{p})$ from $E(\mathbf{p}')$. Thus $E(\mathbf{p}) \cap E(\mathbf{p}') = \emptyset$.

Now we show that, for every one–sided infinite path $\mathbf{p}$ in T, $E(\mathbf{p})$ is uncountable. Without loss of generality we may assume that $\mathbf{p} \in \mathbf{P}_o$. Let $\{x_k\}$ be the sequence of the vertices of $\mathbf{p}$ and let f be the same function as in Example (6.36). Extend f to a function u on T by setting $u(x) = f(x_k)$ if x is a vertex of $T_{x_k} \setminus V(\mathbf{p})$. Then $u \in \mathbf{BD}$ and $\mathbf{p}$ is uncountable since and $u(E(\mathbf{p})) = f(E(\mathbf{p}))$ is uncountable.

Finally we show that, for every path $\mathbf{p}$ and every $z \in E(\mathbf{p})$, the singleton $\{z\}$ is not a G_δ set.

Assume, by way of contradiction, that $\{z\}$ is a G_δ set for some $\mathbf{p}$ and some $z \in E(\mathbf{p})$. By Theorem (6.33) there is $v \in \mathbf{BD}$ such that $v(x) < 1$ on $b\mathfrak{R} \setminus \{z\}$ and $v(z) = 1$. Let $V_n = \{x \in V(\mathbf{p}) : v(x) > 1 - 1/n\}$. Since $\mathbf{p}$ is a recurrent network, we may apply Lemma (6.25) with u the function identically equal to 1 on $V(\mathbf{p})$. Hence there are a function ϕ on $V(\mathbf{p})$ with finite Dirichlet sum and two sequences of vertices $a_m \in V_{2m-1}$ and $b_m \in V_{2m}$ such that $\phi(a_m) = 0$ and $\phi(b_m) = 1$ for all m. Extend ϕ to a function f on T by setting $f(x) = \phi(y)$ for every $y \in V(\mathbf{p})$ and every $x \in T_y \setminus V(\mathbf{p})$. Then $f \in \mathbf{BD}$, $f(a_m) = 0$ and $f(b_m) = 1$. Since a_m and b_m converge to z we have a contradiction.

The above arguments together with (6.38) prove also that Δ has no isolated points.

REMARK. The problem of describing explicitly the Royden boundary of a network seems to be a difficult one, even in the case of "nice" networks. As far as we know the only result in this direction is contained in Wysoczański (1994), where it is proved that the Royden compactification of $\mathbb{N}$ is a quotient subspace of the Cech–Stone compactification $\beta\mathbb{N}$.

§4. The harmonic measure

The existence of Royden's boundary of an infinite network raises the problem of the integral representation of the functions in **HD**. This section is devoted to studying this problem. We will proceed as in Sario and Nakai (1970) in the case of Riemann surfaces.

DEFINITION. Let (Γ, r) be an infinite transient network and let $z \in V$. The harmonic measure on the boundary of (Γ, r) with center at z is a positive regular Borel measure μ_z on $b\Re$ such that every superharmonic function $f \in \mathbf{BD}$ satisfies

$$(6.39) \qquad f(z) \geq \int_{b\Re} f(x) d\mu_z(x).$$

Theorem (6.40). *Let (Γ, r) be a transient network. For every $z \in V$ there exists a unique harmonic measure μ_z on $b\Re$ with center at z. Furthermore, the harmonic measure μ_z has the following properties:*

(1) *$\mu_z(U) > 0$ for all open sets $U \subseteq b\Re$ such that $U \cap \Delta \neq \emptyset$;*

(2) *$\operatorname{supp}(\mu_z) = \Delta$, where $\operatorname{supp}(\mu_z)$ is the support of μ_z;*

(3) *every $f \in \mathbf{HD}$ is μ_z–integrable and $f(z) = \int_\Delta f(x) d\mu_z(x)$;*

(4) *$\mu_z(\Delta) = 1$.*

Proof. First suppose that the harmonic measure with center at z exists. Then for every arbitrary compact nonempty set F in $b\Re \setminus \Delta$ there exists a positive superharmonic function $g \in \mathbf{D}_0$ such that $g = 0$ on Δ and $g = +\infty$ on F (Lemma (6.13) and Theorem (6.12)). Then for every $n > 0$ we have

$$g(z) \geq \min(g, n)(z) \geq \int_{b\Re} \min(g, n) d\mu_z \geq n\mu_z(F).$$

so that $\mu_z(F) = 0$. Hence $\operatorname{supp}(\mu_z) \subseteq \Delta$. Let now U be as in (1). Let F be a compact set in $U \cap \Delta$. Take $f \in \mathbf{BD}$, $0 \leq f \leq 1$ such that $f = 1$ on F and $f = 0$ on $\Delta \setminus U$ (by Lemma (6.6)). Let u be the unique function in **HBD** such that $f - u \in \mathbf{BD}_0$ (by Theorem (6.3)). Then by (6.39) we have

$$u(z) = \int_\Delta u \, d\mu_z = \int_U u \, d\mu_z \leq \mu_z(U).$$

Since $u(z) > 0$ we have (1) and (2). Claim (3) is obvious, when $f \in \mathbf{HBD}$, by (2) and (6.39). We will show below that (3) holds for every $f \in \mathbf{HD}$. Finally, (4) follows from (3), with u the function identically equal to 1.

Now we prove the existence. For every $f \in \mathbf{BD}$ let us denote by $\pi(f)$ the unique function in $\mathbf{HBD}$ such that $f - \pi(f) \in \mathbf{BD}_0$ (i.e., the harmonic component of f in the Royden decomposition of f). By Theorems (6.7) and (6.8)

$$(6.41) \qquad \sup_{z \in \mathfrak{R}} |\pi(f)(x)| = \sup_{z \in \Delta} |f(x)| \le \sup_{z \in \mathfrak{R}} |f(x)|.$$

Since $\mathbf{BD}$ is dense in $C(\mathfrak{R})$ for the uniform topology, the operator π can be extended to $C(\mathfrak{R})$ in such a way that (6.41) is still satisfied. Clearly $\pi(f)$ is bounded and harmonic on V for any $f \in C(\mathfrak{R})$.

Let $C(\Delta)$ denote the space of all continuous real valued functions on the harmonic boundary. By Tietze's extension theorem every function in $C(\Delta)$ can be uniformly approximated by restrictions of functions of $\mathbf{HBD}$.

Let $f \in C(\Delta)$ and let $f' \in C(\mathfrak{R})$ be any extension of f to $\mathfrak{R}$. We set $\rho(f) = \pi(f')$. Such a definition is independent of the choice of f' (e.g. by Theorem (6.7)). We have by the first equation in (6.41)

$$\sup_{z \in \mathfrak{R}} |\rho(f)(x)| = \sup_{z \in \Delta} |f(x)|.$$

Thus the linear functional $f \mapsto \rho(f)(z)$ is bounded on $C(\Delta)$ for every $z \in V$. Hence there exists a positive regular Borel measure μ_z on Δ such that (3) holds for every $f \in \mathbf{HBD}$.

Suppose now $f \in \mathbf{HD}$. By Theorem (3.72) we may assume f positive. Then we can find a sequence of positive functions $f_n \in \mathbf{HBD}$ such that $f_n \le f$ and $f_n(x) \to f(x)$ as $n \to \infty$ (see Theorem (3.73)). Hence (3) holds for every $f \in \mathbf{HD}$.

Now, let f be any superharmonic function in $\mathbf{D}$ and write $f = u + v$ with $u \in \mathbf{HD}$ and $v \in \mathbf{D}_0$. Then v is superharmonic and, by (1) of Lemma (3.70), $v \ge 0$. Hence

$$f(z) \ge u(z) = \int_\Delta u \, d\mu_z = \int_\Delta f \, d\mu_z.$$

Finally we prove the uniqueness of the harmonic measure with center at z. Let μ'_z be another positive regular Borel measure μ_z on $b\mathfrak{R}$ such that (6.39) holds (with μ'_z in place of μ_z) for all superharmonic Dirichlet finite functions. Then μ'_z is supported on the harmonic boundary, as proved above, and defines the same functional as μ_z on $C(\Delta)$. It follows $\mu'_z = \mu_z$. $\square$

Corollary (6.42). *Let x be isolated in Δ. Then $\mu_z(x) > 0$ for every $z \in V$.*

Let o be a fixed reference vertex. In the sequel we will denote simply by μ the harmonic measure with center at o.

Theorem (6.43). *Let (Γ, r) be a transient network. There exists a unique real valued function $K(z, x)$ defined on $V \times \mathfrak{R}$ with the following properties*

(1) $K(z, x)$ *is a nonnegative bounded Borel function on $\mathfrak{R}$ for every fixed $z \in V$;*

(2) $d\mu_z = K(z, \cdot)d\mu$ *for all $z \in V$;*

(3) $K(z, x) = 0$ *on $\Gamma \times \mathfrak{R} \setminus \Delta$;*

(4) $K(o, x) = 1$ *for all $x \in \Delta$;*

(5) $K(z, x)$ *is harmonic on V for μ–almost all $x \in \mathfrak{R}$.*

Proof. For every positive harmonic function u on V and every couple of neighboring vertices $z \sim y$.

$$p(z,y)u(y) \leq \sum_{t \sim z} p(z,t)u(t) = u(z).$$

Exchanging the roles of z and y we get

$$u(y)p(z,y) \leq u(z) \leq u(y)p(y,z)^{-1}.$$

In particular, if $z_0 \sim z_2 \sim \cdots \sim z_n$, then

(6.44) $$u(z_n)\Pi_{j=1}^{n} p(z_{j-1}, z_j) \leq u(z_0) \leq u(z_n)\Pi_{j=1}^{n} p(z_j, z_{j-1})^{-1}.$$

Taking (6.44) into account, we define the discrete Harnack function $k(z_1, z_2)$ for any couple of vertices z_1 and z_2 as

$$k(z_1, z_2) = \inf\{\alpha : \alpha^{-1} \leq \frac{u(z_1)}{u(z_2)} \leq \alpha \text{ for all harmonic functions } u > 0\}.$$

For any positive function $f \in \mathbf{HBD}$, we have by (3) of Theorem (6.40)

$$k(z_1, z_2)^{-1} \int_\Delta f(x)d\mu_{z_2}(x) \leq \int_\Delta f(x)d\mu_{z_1}(x) \leq k(z_1, z_2) \int_\Delta f(x)d\mu_{z_2}(x).$$

By density, we have that μ_z is absolutely continuous with respect to μ. Hence, calling K the Radon–Nikodym derivative

$$K(z, \cdot) = \frac{d\mu_z}{d\mu},$$

we see that $K(z, x)$ satisfies conditions (1) through (4).

For every nonnegative $f \in \mathbf{HBD}$ and every $z \in V$ we have

$$\int_\Delta \left(\sum_{y \sim z} p(z,y)K(y,x)\right) f(x)d\mu(x) = \sum_{y \sim z} p(z,y) \int_\Delta K(y,x)f(x)d\mu(x)$$

(6.45) $$= \sum_{y \sim z} p(z,y)f(y)$$

$$= f(z) = \int_\Delta K(z,x)f(x)d\mu(x).$$

By density we get from (6.45)

$$\int_\Delta P(K(\cdot, x))(z)f(x)d\mu(x) = \int_\Delta K(z,x)f(x)d\mu(x) \quad \text{for all } f \in C(\Delta).$$

Therefore $K(\cdot, x)$ is harmonic for almost all $x \in \mathfrak{R}$.

Suppose that K' is another function with the same properties as K. Set $H = K - K'$. Then $\int_\Delta H(z,x)f(x)d\mu(x) = 0$ for every $z \in V$ and for any continuous function on Δ. Hence there is a set $E_z \subset \Delta$ of μ–measure 0 such that $H(z,x) = 0$ for all $x \notin E_z$. Set $E = \cup_{z \in V} E_z$. Then $\mu(E) = 0$ and $H(z,x) = 0$ for every $z \in V$ and almost all $x \in \Delta \setminus E$. $\square$

DEFINITION. The function $K(z,x)$ of Theorem (6.43) is called the harmonic kernel. Every $f \in \mathbf{HD}$ has the representation

$$(6.46) \qquad f(z) = \int_\Delta K(z,x)f(x)d\mu(x) \quad \text{for all } z \in V.$$

We may ask which functions on V can be represented in the form (6.45) for some μ–integrable f on Δ.

Theorem (6.47). *Let (Γ, r) be a transient network. Let f be a function on Δ integrable with respect to μ and define $f(z)$ as in (6.46) for every $z \in V$. Then f is harmonic on V and there exists a sequence of functions $f_n \in \mathbf{HBD}$ such that $\lim f_n(z) = f(z)$ for all $z \in V$. If in addition f is bounded on Δ and continuous at $x_0 \in \Delta$ as a function on Δ, then $f(z) \to f(x_0)$ as $z \in V$ tends to x_0.*

Proof. On one hand the restrictions of $\mathbf{HBD}$ functions to Δ are dense in $C(\Delta)$ for the uniform topology. On the other hand every $f \in L^1(d\mu)$ can be approximated in the L^1 norm by functions in $C(\Delta)$. Since $\mu(\Delta) = 1$, we conclude that there exists a sequence of functions $f_n \in \mathbf{HBD}$ such that $\int_\Delta |f_n(x) - f(x)|d\mu(x) \to 0$ as $n \to \infty$. Since $f_n(z) = \int_\Delta K(z,x)f_n(x)d\mu(x)$ and $K(z, \cdot)$ is bounded, $f_n(z)$ tends to $f(z)$ for every $z \in V$.

Next, since $K(\cdot, x)$ is harmonic and $K(z, \cdot)$ is bounded,

$$\int_\Delta |f(x)|d\mu(x) \sum_{y \sim z} p(z,y)K(y,x) = \int_\Delta |f(x)||K(z,x)d\mu(x) < \infty.$$

Therefore, by the Tonelli–Fubini theorem, f is harmonic at every $z \in V$.

Finally, fix $\epsilon > 0$. If f is continuous at x_0 as a function on Δ, then there exists a neighborhood $U \subseteq \Delta$ of x_0 such that $|f(x) - f(x_0)| < \epsilon$ for all $x \in U$. If f is bounded, then there exists $u \in \mathbf{HBD}$ such that $u(x_0) = 0$ and $|f(x_0) - f(x)| <$

$u(x) + \epsilon$ for every $x \in \Delta$ (use e.g. Lemma (6.6)). Then we have

$$\left| \int_\Delta K(z,x)f(x)d\mu(x) - f(x_0) \right| \leq \int_\Delta K(z,x)|f(x) - f(x_0)|d\mu(x)$$

$$\leq \int_\Delta K(z,x)(u(x) + \epsilon)d\mu(x)$$

$$= u(z) + \epsilon.$$

It follows that $\limsup |f(z) - f(x_0)| \leq \epsilon$ as $z \in V$ tends to x_0. $\square$

DEFINITION. Let f_1 and f_2 be harmonic functions with finite Dirichlet sums. We denote by $f_1 \vee f_2$ the least harmonic majorant in **HD** and by $f_1 \wedge f_2$ the greatest harmonic minorant in **HD** of f_1 and f_2 i.e.,

$$f_1 \vee f_2(z) = \inf\{u(z) : u \geq f_1, u \geq f_2 \quad u \in \mathbf{HD}\}$$

$$f_1 \wedge f_2(z) = \sup\{u(z) : u \leq f_1, u \leq f_2, \quad u \in \mathbf{HD}\}.$$

For general harmonic functions f_1 and f_2, the greatest harmonic minorant and least harmonic majorant are not necessarily harmonic. However, **HD** is a vector lattice with respect to the operations $\vee$ and $\wedge$.

Theorem (6.48). *Let (Γ, r) be a transient network. Let $f_1, f_2 \in \mathbf{HD}$. Then $f_1 \vee f_2$ and $f_1 \wedge f_2$ are well defined functions in $\mathbf{HD}$. Moreover, for all $z \in V$,*

$$f_1 \vee f_2(z) = \int_\Delta \max(f_1(x), f_2(x))K(z,x)d\mu(x),$$

$$f_1 \wedge f_2(z) = \int_\Delta \min(f_1(x), f_2(x))K(z,x)d\mu(x).$$

Proof. As $f_1 \wedge f_2 = -(-f_1 \vee -f_2)$, we may restrict our considerations to $f_1 \vee f_2$. Moreover, since $f_1 \vee f_2 = f_2 + (f_1 - f_2) \vee 0$, we may confine ourselves to arguing for $f \vee 0$. Set $g(x) = \max(f(x), 0)$. Then $g \in \mathbf{D}$ and g is subharmonic.

There are $u \in \mathbf{HD}$ and $v \in \mathbf{D}_0$ such that $g = u + v$. We have $v \leq 0$ by the Remark to Lemma (3.70). Therefore $u \geq f$ and $u \geq 0$. Let $w \in \mathbf{HD}$ be such that $w \geq f$ and $w \geq 0$. Then $w \geq u$ on Δ, whence $w \geq u$ on V. It follows $u = f \vee 0$.

By (3) of Theorem (6.40), for all $z \in V$,

$$u(z) = \int_\Delta u(x)K(z,x)d\mu(x) = \int_\Delta g(x)K(z,x)d\mu(x)$$

thus concluding the proof.

§5. Spaces HD of finite dimension

DEFINITION. For every integer $n > 0$ we denote by $\mathcal{O}_{\mathbf{HD}}^n$ the class of all transient networks (Γ, r) such that $\dim \mathbf{HD} = n$.

With this definition $\mathcal{O}_{\mathbf{HD}} = \mathcal{O}_{\mathbf{HD}}^1$. For completeness we say that a recurrent network belongs to $\mathcal{O}_{\mathbf{HD}}^0$. Note that, by Theorem (3.73), $\dim \mathbf{HD} = \dim \mathbf{HBD}$.

Lemma (6.49). *A transient network (Γ, r) belongs to $\mathcal{O}_{\mathbf{HD}}^n$ $(1 \le n < \infty)$ if and only if Δ consists of n points. In this case $\mathbf{HD} = \mathbf{HBD}$.*

Proof. Suppose first that (Γ, r) is in $\mathcal{O}_{\mathbf{HD}}^n$ and that Δ contains at least $n + 1$ distinct points $x_1, \ldots, x_{n+1}$. We can find pairwise disjoint neighborhoods U_j of x_j and functions $f_j \in \mathbf{BD}$ such that $f_j(x_j) = 1$, $0 \le f_j \le 1$ and $f_j = 0$ on $\mathfrak{R} \setminus U_j$.

Let $f_j = \phi_j + v_j$ be the Royden decomposition of f_j, with $\phi_j \in \mathbf{HBD}$ and $v_j \in \mathbf{BD}_0$. Suppose that

$$(6.50) \qquad \phi = \kappa_1 \phi_1 + \kappa_2 \phi_2 + \cdots \kappa_{n+1} \phi_{n+1} = 0 \quad \text{on } V$$

for some coefficients $\kappa_1, \ldots, \kappa_{n+1}$. Then $\phi = 0$ on Δ. Since $\phi(x_j) = \kappa_j$, we have that all the coefficients in (6.50) must be zero. This contradicts $\dim \mathbf{HD} = n$.

Thus Δ consists of $m \le n$ points $x_1, \ldots, x_m$. By Theorem (6.20) for every $j = 1, \ldots, m$ there is a $\mathbf{HD}$–minimal harmonic function u_j such that $u_j(x_i) = 0$ if $i \ne j$ and $u_j(x_j) = 1$. Let u be any function in $\mathbf{HD}$. Then

$$u(x) = \sum_{j=1}^m u(x_j) u_j(x) \quad \text{for all } x \in \mathfrak{R}.$$

Hence the $\mathbf{HD}$–minimal harmonic functions u_j form a basis for $\mathbf{HD} = \mathbf{HBD}$. Thus $n = m$. It is clear that these arguments prove also the opposite implication. $\square$

Corollary (6.51). *A transient network (Γ, r) belongs to $\mathcal{O}_{\mathbf{HD}}$ if and only if Δ is a singleton.*

We saw in §2 of Chapter IV that network automorphisms are of some interest in the study of the class $\mathcal{O}_{\mathbf{HD}}^1$. As a further result in this direction, now we will prove that a necessary condition for a locally finite graph to be in $\cup_{n=2}^{\infty} \mathcal{O}_{\mathbf{HD}}^n$ is the compactness of the automorphisms group $\mathrm{Aut}(\Gamma)$. No continuous counterpart of this result is so far known.

Suppose that Γ is a locally finite graph with all the resistances equal to 1. Following Trofimov (1985)(b) we define a metric ρ on $\mathrm{Aut}(\Gamma)$ in the following way.

Fix $o \in V$ and let $B(o, n) \subset V$ be the ball of radius n centered at o. For every $\phi \in \mathrm{Aut}(\Gamma)$ let $\mathrm{Fix}(\phi) \subset V$ denote the set of all $x \in V$ such that $\phi(x) = x$. We set, for all $\phi_1, \phi_2 \in \text{Aut}(\Gamma)$

$$\rho(\phi_1, \phi_2) = \begin{cases} 0 & \text{if } \phi_1 = \phi_2; \\ n & \text{if } d(o, \phi_1^{-1}\phi_2(o)) = n; \\ 2^{-n-1} & \text{if } B(o,n) \subseteq \text{Fix}(\phi_1^{-1}\phi_2) \text{ and } B(o, n+1) \not\subseteq \text{Fix}(\phi_1^{-1}\phi_2). \end{cases}$$

It is not difficult to verify that $\text{Aut}(\Gamma)$ becomes a Hausdorff locally compact group and that the subgroups

(6.52) $H_n = \{g \in \text{Aut}(\Gamma) : \rho(\phi, o) \le 2^{-n-1}\}, \quad -\infty < n < +\infty$

form a basis of compact open neighborhoods of the identity. In particular, a subset H is compact if and only if H is closed and bounded.

Theorem (6.53). *Let $\Gamma = (V, Y)$ be an infinite locally finite graph. Let $\text{Aut}(\Gamma)$ be the automorphisms group of Γ. The following are equivalent*

(1) *$\text{Aut}(\Gamma)$ is compact;*

(2) *there exists a finite set $F \subset V$ such that $g(F) = F$ for all $g \in \text{Aut}(\Gamma)$;*

(3) *for some $(=$all$)$ $x \in V$ the set $O(x) = \{\phi(x) : \phi \in \text{Aut}(\Gamma)\}$ is finite.*

Proof. Assume that (1) holds. Let H be the stabilizer of o and let $\{gH\}$ be the left cosets space of H. Since H is open there are finitely many cosets $\phi_1 H, \dots, \phi_m H$ whose union is equal to $\text{Aut}(\Gamma)$. It is easily seen that the set $F = \{\phi_1(o), \dots, \phi_m(o)\}$ satisfies (2).

A fortiori (2) implies (3). Finally, assume that (3) holds for some $x \in V$. Then $O(z)$ is finite for all $z \in V$. In particular $O(o)$ is finite. Hence there exists n such that $\text{Aut}(\Gamma) \subseteq H_n$ i.e., $\text{Aut}(\Gamma)$ is compact. $\square$

Theorem (6.54). *Let (Γ, r) be a transient locally finite network with all the resistances equal to 1. Suppose that $2 \le \dim \mathbf{HD} < \infty$. Then $\text{Aut}(\Gamma)$ is compact.*

Proof. Let $\phi : V \mapsto V$ be an automorphism of Γ. For every $f \in \mathbf{BD}$ let $f \circ \phi(x) = f(\phi(x))$, $x \in V$. Clearly $f \circ \phi \in \mathbf{BD}$. Moreover, $f \circ \phi$ belongs to $\mathbf{HBD}$ if f does.

For every $y \in b\Re$ we define $\phi(y)$ as the multiplicative functional $f \mapsto f \circ \phi(y)$. If $\{x_\alpha\}$ is a net in $\Re$ such that $\lim_\alpha x_\alpha = y \in \Re$, then $f(x_\alpha) \to f(y)$ for all $f \in \mathbf{BD}$. Hence $f \circ \phi(x_\alpha) \to f \circ \phi(y)$. Thus ϕ can be extended to a homeomorphism of $\Re$. Since $f \circ \phi \in \mathbf{BD}_0$ whenever $f \in \mathbf{BD}_0$, we see that $\phi(\Delta) = \Delta$.

For every $y \in \Delta$, let $H(y)$ be the stabilizer of y i.e.,

$$H(y) = \{\phi \in \text{Aut}(\Gamma) : \phi(y) = y\}$$

We will prove that $H(y)$ is compact. As already noticed, this amounts to proving that $H(y)$ is closed and bounded.

First we show that $H(y)$ is closed. Assume that $\phi_n \in H(y)$ for $n = 1, 2, \ldots$ and let ϕ_n tend to ϕ as $n \to \infty$. By Lemma (6.49) y is an isolated point in Δ, so that there exists a **HD**–minimal harmonic function u such that $u(y) = 1$ and $u(x) = 0$ for every $x \in \Delta \setminus \{y\}$. Set $f = u \circ \phi$ and $f_n = u \circ \phi_n$. Since $\phi(\Delta) = \Delta$, f and f_n are **HD**–minimal harmonic functions as well. Since $\phi_n \in H(y)$, $f_n = u$ for all n. On the other hand, if $\phi \notin H(y)$, then $f \neq u$. Hence there exists $z \in V$ such that $f(z) \neq u(z)$. Since $\phi_n \to \phi$ in $\mathrm{Aut}(\Gamma)$, $\phi_n(z) = \phi(z)$ for large values of n. So

$$u(z) = u(\phi_n(z)) = u(\phi(z)) \quad \text{if } n \text{ is large,}$$

a contradiction. Hence $\phi \in H(y)$ and $H(y)$ is closed.

Next we show that $H(y)$ is bounded. Assume, by contradiction, that $H(y)$ is unbounded. Then there is a sequence of automorphisms $\phi_n \in H(y)$ such that $d(o, \phi_n(o)) \to \infty$ as $n \to \infty$. By connectedness

$$(6.55) \qquad\qquad d(o, \phi_n(x)) \to \infty \quad \text{for all } x \in V.$$

As above, let u be the **HD**–minimal positive harmonic function such that $u(y) = 1$ and $u(x) = 0$ for every $x \in \Delta \setminus \{y\}$. We still have $u \circ \phi_n = u$ for all n.

By (6.55) and since $D(u) < \infty$

$$0 = \lim_{n \to \infty} \left(u(\phi_n(x)) - u(\phi_n(z)) \right) = u(x) - u(z) \quad \text{for every } z \sim x \in V.$$

By connectedness $u(x) = u(z)$ for any two vertices in V. Thus u is constant, so that Δ reduces to the singleton $\{y\}$. By Lemma (6.49) $\dim \mathbf{HD} = 1$, contrary to our assumption. Hence $H(y)$ is bounded. This proves our claim that $H(y)$ is compact.

According to Lemma (6.49), $\Delta = \{x_1, \ldots, x_n\}$ for some $n \geq 2$. Let $H = H(x_1)$ be the stabilizer of x_1. Let $O(x_1) = \{\phi(x_1) : \phi \in \mathrm{Aut}(\Gamma)\}$. Then

$$O(x_1) = \{x_{j_1}, \ldots, x_{j_k}\} \subseteq \Delta.$$

Hence there are k automorphisms $\phi_1, \ldots, \phi_k$ such that $\phi_1(x_1) = x_{j_1}, \ldots, \phi_k(x_1) = x_{j_k}$. Denoting by ϕH a left coset of H, we have

$$\mathrm{Aut}(\Gamma) = \phi_1 H \cup \phi_2 H \cup \cdots \phi_k H.$$

Since H is compact we have the thesis. $\square$

EXAMPLE (6.56). The compactness of $\mathrm{Aut}(\Gamma)$ is by no means sufficient to prove that $\dim \mathbf{HD}$ is finite and greater than 1. For instance, let $\Gamma = T$ be the infinite binary tree with root o. By this we mean that o has degree 2 and every other vertex has degree 3. Clearly every automorphism of T must fix o. Therefore $\text{Aut}(\Gamma)$ is compact by Theorem (6.53).

By (3.58) we get that the resistance between o and infinity is 1. Hence the tree is transient. Now, for every $x \in V$, let T_x be the unique binary subtree with vertex x. By Corollary (4.25) there exists a noncostant harmonic function u_x on T_x such that $u_x(x) = 0$ and $D(u_x) < \infty$. Extend u_x to the whole of T by defining u_x to be zero outside T_x. Then u_x is harmonic on T and has finite Dirichlet sum. Thus $\mathbf{HD}$ is not finite dimensional.

§6. Compactification of subnetworks

Let (Γ, r) be an infinite network, with $\Gamma = (V, Y)$. Let (Υ, r) be an infinite subnetwork of (Γ, r), with $\Upsilon = (U, E)$.

In the sequel, when dealing with subnetworks (Υ, r) of a network (Γ, r), we will use the following notation: $\mathbf{BD}(\Gamma)$ and $\mathbf{BD}(\Upsilon)$ will denote the Dirichlet algebras of the networks (Γ, r) and (Υ, r), respectively. The notation $\mathbf{BD}_0(\Gamma)$, $\mathbf{BD}_0(\Upsilon)$, $\mathbf{HBD}(\Upsilon)$ etc. will have the analogous meaning. Moreover $\mathbf{BD}(\Gamma; \Upsilon)$, $\mathbf{BD}_0(\Gamma; \Upsilon)$, etc. will be the space of the restrictions to U of the functions in $\mathbf{BD}(\Gamma)$, $\mathbf{BD}_0(\Gamma)$, etc. Note that $\mathbf{BD}(\Gamma; \Upsilon)$ is a subalgebra of $\mathbf{BD}(\Upsilon)$ and $\mathbf{BD}_0(\Gamma; \Upsilon)$ is a subalgebra of $\mathbf{BD}_0(\Upsilon)$.

The Royden compactification and the boundaries of (Γ, r) will be denoted as before by $\mathfrak{R}$, $b\mathfrak{R}$ and Δ. The Royden compactification of (Υ, r) will be denoted by $\mathfrak{R}_\Upsilon$, while the Royden and the harmonic boundary will be denoted by $b\mathfrak{R}_\Upsilon$ and Δ_Υ respectively. The closure in $\mathfrak{R}$ of U will be denoted by $\overline{U}$. The fundamental relation between the $\overline{U}$ and $\mathfrak{R}_\Upsilon$ is provided by the following theorem.

Theorem (6.57). *Let (Υ, r) be an infinite subnetwork of (Γ, r). Then there exists a unique continuous mapping Θ of $\mathfrak{R}_\Upsilon$ onto $\overline{U}$ fixing U elementwise. Furthermore, Θ maps Δ_Υ to $\Delta \cap \overline{U}$.*

Proof. The proof is the same as in the case of Riemann surfaces. Uniqueness is obvious by the density of U both in $\overline{U}$ and $\mathfrak{R}_\Upsilon$. To prove existence first notice that a function $f \in \mathbf{BD}(\Gamma; \Upsilon)$ can be extended by continuity both to a function on $\mathfrak{R}_\Upsilon$ and to a function on $\overline{U}$. Denote by f^* and $\overline{f}$ respectively such extensions and choose any $x^* \in \mathfrak{R}_\Upsilon$. The map $f \mapsto f^*(x^*)$ defines a continuous character on $\mathbf{BD}(\Gamma; \Upsilon)$. We claim that there exists a unique $\overline{x} \in \overline{U}$ such that

$$(6.58) \qquad \overline{f}(\overline{x}) = f^*(x^*) \quad \text{for all } f \in \mathbf{BD}(\Gamma; \Upsilon).$$

Suppose on the contrary that for every $y \in \overline{U}$ there exists $g_y \in \mathbf{BD}(\Gamma; \Upsilon)$ such that $\overline{g}_y(y) \neq 0$ and $g_y^*(x^*) = 0$. We may assume g_y positive and $\overline{g}_y(y) > 1$. By the compacteness of $\overline{U}$ there are points $y_1, \ldots, y_n$ such that $\overline{v} = \sum_{j=1}^{n} \overline{g}_{y_n} \geq 1$ on $\overline{U}$.

We have $v^*(x^*) = 0$ and $v \geq 1$ on U, a contradiction. Hence there exists $\overline{x} \in \overline{U}$ such that $g^*(x^*) = 0$ implies $\overline{g}(\overline{x}) = 0$ for all $g \in \mathbf{BD}(\Gamma; \Upsilon)$. Letting $g = f - f^*(x^*)$ we obtain (6.58). Clearly, for every x^* there is a unique $\overline{x}$ because $\mathbf{BD}(\Gamma; \Upsilon)$ separates points in $\overline{U}$.

If x^* and $\overline{x}$ are as above we set

$$(6.59) \qquad\qquad \overline{x} = \Theta(x^*).$$

We have just shown that

$$(6.60) \qquad \overline{f}(\Theta(x^*)) = f^*(x^*) \quad \text{for all } f \in \mathbf{BD}(\Gamma; \Upsilon).$$

Continuity of Θ follows from (6.60). It is also clear that Θ fixes U elementwise.

Now we have to prove that Θ is onto. Let $\overline{z} \in \overline{U}$ and denote by $\mathbf{I}$ the ideal of all functions $f \in \mathbf{BD}(\Upsilon)$ such that $f(x) \to 0$ as $x \in U$ tends to $\overline{z}$. Clearly $\mathbf{I}$ is a proper ideal since it does not contain the function identically 1 on U. Therefore $\mathbf{I}$ is contained in a maximal ideal $\mathbf{M}$. As a consequence there exists a multiplicative linear functional x^* on $\mathbf{BD}(\Upsilon)$ which annhilates all the elements of $\mathbf{M}$ (see Yosida (1970) or Larsen (1973)). By the definition of Royden's compactification (see Theorem (6.4)), $x^* \in \mathfrak{R}_\Upsilon$ and $f^*(x^*) = 0$ for all $f \in \mathbf{I}$. Now, if f is arbitrary in $\mathbf{BD}(\Gamma; \Upsilon)$, then $f - \overline{f}(\overline{z})$ is in $\mathbf{I}$ so that $f^*(x^*) = \overline{f}(\overline{z})$.

Finally, let x^* belong to Δ_Υ. Let $\overline{x}$ be as in (6.59). By (6.60) $\overline{f}(\Theta(x^*)) = 0$ for every $f \in \mathbf{BD}_0(\Gamma; \Upsilon)$. Therefore $\overline{f}(\overline{x}) = 0$ for all $f \in \mathbf{BD}_0(\Gamma)$. $\square$

REMARK. There are two "boundaries" in $\overline{U}$. The first one is the closure $\overline{\daleth U}$ of the combinatorial boundary of U. The second boundary is

$$b_U = (\overline{U} \setminus \overline{\daleth U}) \cap b\mathfrak{R}.$$

Accordingly, the "boundary" of $\mathfrak{R}_\Upsilon$ splits into two parts, $\Theta^{-1}(\overline{\daleth U})$ and $\Theta^{-1}(b_U)$. The structure of $\overline{\daleth U}$ is different from the structure of $\Theta^{-1}(\overline{\daleth U})$. In fact a function $f \in \mathbf{D}(\Upsilon)$ is continuous at each point $\Theta^{-1}(\overline{\daleth U})$ but it may be discontinuous on $\overline{\daleth U}$. On the contrary, the structures of b_U and $\Theta^{-1}(b_U)$ are the same, provided that the graph Υ is induced by U. We start with a lemma.

Lemma (6.61). Let (Γ, r) be an infinite network and let U be an infinite connected subset of vertices. Then $b_U \cup U$ is open in $\mathfrak{R}$.

Proof. Suppose that b_U is not empty and let $z \in b_U$. Then there exists an open neighborhood $A \subseteq \mathfrak{R}$ of z such that $\overline{A}$ does not contain points of $\daleth U$. Analogously, there exists an open set B such that $\overline{A} \cap \overline{B} = \emptyset$ and $\neg U \subseteq B$. Therefore there exists a function $f \in \mathbf{BD}(\Gamma)$ such that $0 \leq f \leq 1$, $f = 1$ on A and $f = 0$ on B. In particular $f = 0$ on $\neg U$. Define a new function g on V in the following way

$$g(x) = f(x) \quad \text{if } x \in U$$
$$g(x) = 0 \qquad \text{if } x \notin U.$$

We claim that $g \in \mathbf{BD}(\Gamma)$. Actually, let $x \sim y$ be vertices in Γ. If both nodes x and y are in U, then $g(x) - g(y) = f(x) - f(y)$. If $x \in U$ but $y \notin U$, then $x \in \neg U$ so that $g(x) = f(x) = 0 = g(y)$. If both x and y are not in U then $g(x) = g(y) = 0$. Therefore $D(g) \leq D(f) \leq \infty$. As $g(z) = f(z) = 1$ and $g(x) = 0$ for $x \in \Re \setminus (b_U \cup U)$ we may conclude that z is an interior point of $b_U \cup U$. $\square$

Now we are ready to prove the equivalence mentioned above.

Theorem (6.62). *Let (Υ, r) be an infinite subnetwork of (Γ, r) such that Υ is induced by its vertex set U. Let Θ be the mapping of Theorem (6.57). Then Θ is a homeomorphism of $U \cup \Theta^{-1}(b_U)$ onto $U \cup b_U$. Furthermore, $\Theta^{-1}(\overline{x}) \in \Delta_\Upsilon$ for every $\overline{x} \in b_U \cap \Delta$.*

Proof. Assume that there exists $\overline{x} \in b_U$ such that $\Theta^{-1}(\overline{x})$ contains two distinct points x_1^* and x_2^* (we denote by f^* the extension of f to $\Re_\Upsilon$). Let $f \in \mathbf{BD}(\Upsilon)$ be such that $f^*(x_j^*) = j$ for $j = 1, 2$. By Lemma (6.61) there exists an open neighborhood $A \subseteq U \cup b_U$ of $\overline{x}$ and an open set B such that $\overline{A} \cap \overline{B} = \emptyset$ and $(\Re \setminus \overline{U}) \cup \neg U \subset B$. Let $g \in \mathbf{BD}(\Gamma)$ be a function such that $g = 1$ on A and $g = 0$ on B. In particular, $g = 0$ on $\neg \Upsilon$. Since Υ is induced by U, the product $h = fg$ can be extended to a function belonging to $\mathbf{BD}(\Gamma)$ by setting h equal to zero on $V \setminus U$.

Thus we have for $j = 1, 2$

$$j = f^*(x_j)g^*(x_j^*) = (fg)^*(x_j^*)$$
$$= \overline{fg}(\Theta(x_j^*)) = \overline{fg}(\overline{x}),$$

which is absurd. Therefore $\Theta^{-1}(\overline{x})$ is a singleton for every $\overline{x} \in U \cup bU$. Since Θ is continuous and surjective and $\Re_\Upsilon$ is compact, Θ maps open subsets of $\Theta^{-1}(b_U)$ to open subsets of b_U. Taking Lemma (6.61) into account, we may conclude that Θ is a homeomorphism of $U \cup \Theta^{-1}(b_U)$ onto $U \cup b_U$.

Let $\overline{x} \in b_U$ be a harmonic boundary point of (Γ, r). Let f be any function in $\mathbf{BD}_0(\Upsilon)$ and define $g \in \mathbf{BD}(\Gamma)$ as above. Arguing as above we have that $fg \in \mathbf{BD}(\Gamma; \Upsilon)$. In fact it is easy to check that $fg \in \mathbf{BD}_0(\Gamma; \Upsilon)$. Let us denote by

h a function in $\mathbf{BD}_0(\Gamma)$ which extends fg to V. We have

$$f^*(\Theta^{-1}(\overline{x})) = (fg)^*(\Theta^{-1}(\overline{x}))$$
$$= \overline{fg}(\overline{x}) = h(\overline{x}) = 0.$$

Since it annhilates every function in $\mathbf{BD}_0(\Upsilon)$, $\Theta^{-1}(\overline{x})$ is a harmonic boundary point of (Υ, r). $\square$

As a consequence of Theorem (6.62) we can make more precise the result in Theorem (4.20). We keep the notation introduced in §4 of Chapter IV.

Corollary (6.63). *Let (Γ, r) be an infinite network with the following property: there exists a finite subset of vertices L such that $\Gamma(L)$ has exactly $n \geq 2$ infinite connected components $\Gamma_j(L)$, $j = 1, \ldots, n$, all of whose are transient and belong to $\mathcal{O}_{\mathbf{HD}}$. Then $(\Gamma, r) \in \mathcal{O}_{\mathbf{HD}}^n$.*

Proof. Set $\Gamma_j(L) = \Upsilon_j$ and denote by U_j the vertex set of Υ_j. The function which is 1 on U_j, and 0 on U_i for every $i \neq j$, belongs to $\mathbf{BD}(\Gamma)$. It follows that

$$\overline{U_j} \cap \overline{U_i} = \emptyset \quad \text{for all } i \neq j.$$

Therefore we have

(6.64) $U_j \cup b_{U_j} = \overline{U_j} \quad \text{for all } j = 1, \ldots, n.$

Let $\Theta_j : \mathfrak{R}_{\Upsilon_j} \mapsto \overline{U_j}$ be the map of Theorem (6.57). By Theorem (6.62) Θ_j is a homeomorphism for all j. Hence we have, both set theoretically and topologically,

(6.65) $$b\mathfrak{R} = \bigcup_{j=1}^{n} b_{U_j}.$$

Since $\Upsilon_j \in \mathcal{O}_{\mathbf{HD}}$, Lemma (6.49) implies that the harmonic boundary of (Υ_j, r) consists of exactly one point. By Theorem (6.62) each b_{U_j} contains exactly one harmonic boundary point of (Γ, r). By (6.64) and (6.65) the harmonic boundary of (Γ, r) consists of n points. Therefore $(\Gamma, r) \in \mathcal{O}_{\mathbf{HD}}^n$ by Lemma (6.49) again. $\square$

The last result of this section is the discrete version of a theorem due to Kusunoki and Mori (1959), (1960) for Riemann surfaces (see also Grigor'yan (1988)). However, the proof is somewhat different. We start with the definition of the class $\mathcal{SO}_{\mathbf{HD}}$.

DEFINITION. Let (Γ, r) be a network. We will say that a subnetwork (Υ, r) belongs to the class $\mathcal{SO}_{\mathbf{HD}}$ if there is no nontrivial function $f \in \mathbf{BD}(\Upsilon)$ which is harmonic on $U \setminus \daleth U$ and vanishes on $\daleth U$.

Lemma (6.66). *Let* (Υ, r) *be an infinite subnetwork of a network* (Γ, r). *Suppose that* Υ *is induced by its vertex set* U. *Then* (Υ, r) *does not belong to* SO_{HD} *if and only if* $b_U \cap \Delta \neq \emptyset$.

Proof. Suppose first that there exists a nontrivial $u \in \mathbf{BD}(\Upsilon)$ which is harmonic on $U \setminus \mathbb{T}U$ and vanish on $\mathbb{T}U$. Then u^2 is subharmonic. In fact if $x \in \mathbb{T}U$

$$u^2(x) = 0 \leq \sum_{y \sim x} p(x,y) u^2(y),$$

while, for $x \in U \setminus \mathbb{T}U$,

$$u^2(x) = \left(\sum_{y \sim x} p(x,y) u(y) \right)^2 \leq \sum_{y \sim x} p(x,y) u^2(y).$$

Since $u^2 \in \mathbf{BD}(\Upsilon)$, (Υ, r) is transient by Theorem (3.34). Let $z \in U$ be such that $u^2(z) > 0$. Then we have by (6.39)

$$-u^2(z) \geq \int_{\Delta_\Upsilon} -u^2(x) d\nu_z(x),$$

where ν_z is the harmonic measure with center at z on the harmonic boundary of (Υ, r). Therefore there must exist $x^* \in \Delta_\Upsilon$ such that $u(x^*) \neq 0$. By Theorem (6.57) $\bar{x} = \Theta(x^*) \in \Delta \cap \overline{U}$. We may extend the function u^2 to the whole of V by setting $u^2 = 0$ on $V \setminus U$. This extension belongs to $\mathbf{BD}(\Gamma)$ (as Υ is induced by U) and separates $\Theta(x^*)$ from $\mathbb{T}U$. Therefore $\Theta(x^*) \in b_U$.

Conversely, let $\bar{x} \in b_U \cap \Delta$. Then there exists $g \in \mathbf{BD}(\Gamma)$ such that $g(\bar{x}) \neq 0$ and $g = 0$ on $\mathbb{T}U$. Let f denote the restriction of g to U. Then

$$(6.67) \qquad\qquad f(\Theta^{-1}(\bar{x})) \neq 0$$

by (6.60). Hence f is not in $\mathbf{BD}_0(\Upsilon)$. By applying Theorem (6.3) with $F = \mathbb{T}U$ we get that there are $v \in \mathbf{BD}_0(\Upsilon)$, $v = 0$ on $\mathbb{T}U$, and $u \in \mathbf{BD}(\Upsilon)$, harmonic on $U \setminus \mathbb{T}U$, such that $f = u + v$. Note that u is not identically zero by (6.67). Since $f = 0$ on $\mathbb{T}U$ we have $u = 0$ on $\mathbb{T}U$. Therefore (Υ, r) is not in the class SO_{HD}. $\square$

Finally we come to Kusunoki–Mori theorem for networks.

Theorem (6.68). *Let* (Γ, r) *be an infinite network such that* $\sup_{x \sim y} r(x, y) < \infty$. *Let* m *be any nonnegative integer. Then* $\dim \mathbf{HD} > m$ *if and only if there exist* $m + 1$ *infinite subnetworks* $(\Upsilon_1, r), \ldots, (\Upsilon_{m+1}, r)$ *with the following properties*

(1) Υ_j *is induced by its vertex set* U_j, *for* $j = 1, \ldots, m + 1$;

(2) $U_j \cap U_i = \mathbb{T}U_j \cap \mathbb{T}U_i$ *for all* $j \neq i$;

(3) (Υ_j, r) *is not in the class* SO_{HD} *for* $j = 1, \ldots, m + 1$.

Proof. Suppose first that $m+1$ subnetworks of this kind exist. By Lemma (6.66) $b_{U_j} \cap \Delta \neq \emptyset$ for all j. Let $\bar{x}_j$ be a point $b_{U_j} \cap \Delta$. We claim that $\bar{x}_j \neq \bar{x}_i$ if $i \neq j$. To see this, let $f_j \in \mathbf{BD}(\Upsilon_j)$ be such that $f_j = 0$ on $\daleth U_j$ and $f_j(\bar{x}_j) = 1$. Then, defining g_j as $g_j = f_j$ on U_j and $g_j = 0$ on the complement of U_j we have that $g_j \in \mathbf{BD}(\Gamma)$. By (2) $g_j(\bar{x}_j) = 1$ and $g_j(\bar{x}_i) = 0$ if $i \neq j$. Hence Δ contains at least $m+1$ distinct points. By Lemma (6.49) $\dim \mathbf{HD} > m$.

Conversely assume that $\dim \mathbf{HD} > m$. Always by Lemma (6.49) there are at least $m+1$ distinct points $x_1, \ldots, x_{m+1}$ in Δ. We can find $m+1$ neighborhoods $N_1, \ldots, N_{m+1}$ of $x_1, \ldots, x_{m+1}$ such that $N_j \cap N_i = \emptyset$ for $i \neq j$. Hence there are $m+1$ functions $g_j \in \mathbf{BD}(\Gamma)$ such that $g_j(x_j) = 1$, $0 \leq g_j \leq 1$ and $g_j = 0$ outside N_j. Clearly $g_j \notin \mathbf{BD}_0(\Gamma)$.

Let $V_j = N_j \cap V$. By the assumption on r we may assume that $d(x, y) > 1$ whenever $x \in V_j$ and $y \in V_i$ with $j \neq i$. Note also that V_j is not necessarily connected. In any case $g_j(x) = 0$ if $x \notin V_j$. We use Theorem (6.3) to decompose g_j as a sum $u_j + v_j$, where $v_j \in \mathbf{BD}_0(\Gamma)$ is zero on the complement of V_j and $u_j \in \mathbf{BD}(\Gamma)$ is harmonic in V_j. In particular, $u_j(x) = 0$ if $x \notin V_j$. Since g_j does not belong to $\mathbf{BD}_0(\Gamma)$, u_j is not identically zero.

As already shown in the proof of Lemma (6.66), the function $s_j = u_j^2$ is subharmonic. Let C be a finite connected component of V_j and let

$$C' = \bigcup_{x \in C} V(x).$$

Let Γ' be the connected subgraph induced by C'. The functions s_j is subharmonic on the network (Γ', r) and takes only finitely many distinct values on C'. Therefore s_j is constant on C'. As $s_j(x) = 0$ for $x \in C' \setminus C$, we conclude that $s_j = 0 = u_j$ on C'. Thus, for every $j = 1, \ldots, m+1$, there must be an infinite connected component Y_j of V_j such that u_j is not identically 0 on Y_j. We set

$$U_j = \bigcup_{x \in Y_j} V(x) \quad j = 1, \ldots, m+1,$$

and denote by Υ_j the graph induced by U_j. The subnetworks (Υ_j, r) satisfy conditions (1) and (2).

To prove condition (3) we cannot use the functions u_j, since a point in $U_j \setminus Y_j$ could be an interior point. However, it is true that $\daleth U_j \subseteq U_j \setminus Y_j$. Therefore $u_j = 0$ on $\daleth U_j$. We apply once again Theorem (6.3) with $F = \daleth U_j$. Hence we obtain a function $\tilde{u}_j \in \mathbf{BD}(\Upsilon)$ which vanishes on $\daleth U_j$ and is harmonic on $U_j \setminus \daleth U_j$. $\square$

CHAPTER VII

ROUGH ISOMETRIES

§1. Rough isometries in metric spaces

In this chapter we study an equivalence relation between graphs introduced by Kanai (1985) and Gromov (1981), (1987) for general metric spaces. We are interested in this relation since it preserves several important properties studied in the previous chapters.

DEFINITION. Suppose that (X_1, d_1) and (X_2, d_2) are metric spaces. A map $\phi : X_1 \mapsto X_2$ is called a rough isometry (a quasi isometry in Gromov's language) if the two following conditions are satisfied:

there are positive $a > 0$ and $b \geq 0$ such that, for all $x, y \in X_1$,

$$(7.1) \qquad a^{-1} d_1(x, y) - b \leq d_2(\phi(x), \phi(y)) \leq a d_1(x, y) + b;$$

there is a constant $c > 0$ such that, for every $z \in X_2$, there exists $x \in X_1$ such that

$$(7.2) \qquad d_2(z, \phi(x)) \leq c.$$

We will call a, b and c the constants of the rough isometry. If X_1 and X_2 are metric spaces as above, then we will say that X_1 and X_2 are roughly isometric. This terminology is motivated by Theorem (7.3) below.

Note that a rough isometry in general is not continuous nor surjective. It is also clear that the definition of rough isometry is meaningful only in spaces with infinite diameter.

Theorem (7.3). *To be roughly isometric is an equivalence relation among metric spaces.*

Proof. Suppose that $\phi : X \mapsto Y$ is a rough isometry with constant a_1, b_1 and c_1 and that $\psi : Y \mapsto Z$ is a rough isometry with constant a_2, b_2 and c_2. Then the composition $\xi = \psi \circ \phi : X \mapsto Z$ is a rough isometry with constants $a = a_1 a_2$, $b = a_2 b_1 + b_2$ and $c = a_2 c_1 + b_2 + c_2$. Hence the relation is transitive.

Now we prove simmetry. Let $\phi : X_1 \mapsto X_2$ be a rough isometry. We define a map $\overline{\phi} : X_2 \mapsto X_1$ in the following way. For every $z \in X_2$ choose any $x \in X_1$ satisfying (7.2). We put $\overline{\phi}(z) = x$. The map $\overline{\phi}$ is well defined and

$$a^{-1}d_2(z,t) - a^{-1}(b+2c) \le d_1(\overline{\phi}(z),\overline{\phi}(t)) \le ad_2(z,t) + a(b+2c).$$

To prove that the map $\overline{\phi}$ satisfies (7.2) choose any $x \in X_1$ and set $z = \phi(x)$. Then there is some $x' \in X_1$ such that $d_2(\phi(x'),z) \le c$ and $\overline{\phi}(z) = x'$. Clearly $d_1(x,x') \le a(c+b)$. This concludes the proof. $\square$

REMARK. The map $\overline{\phi} : X_2 \mapsto X_1$ defined in the proof of Theorem (7.3) is called a rough inverse of ϕ. Both $d_1(\overline{\phi}(\phi(x)),x)$ and $d_2(\phi(\overline{\phi}(z)),z)$ are bounded in X_1 and in X_2, respectively.

Kanai (1986)(a) proved that if X_1 and X_2 are roughly isometric, connected Riemannian manifolds of bounded geometry, then X_2 is parabolic if X_1 is. This result was obtained by showing that a Riemannian manifold is parabolic if and only if certain roughly isometric graphs (ϵ-nets) defined on the manifold are recurrent. This approach was also taken by Markvorsen, Mc Guinness and Thomassen (1992) to prove that Scherk's surface is hyperbolic.

From now on we will deal only with rough isometries between graphs. For extensions to more general networks we refer to Medolla (1994) and to Coulhon and Saloff-Coste (1994). Let us give an explicit definition.

DEFINITION. We will say that two locally finite graphs $\Gamma_1 = (V_1, Y_1)$ and $\Gamma_2 = (V_2, Y_2)$ are roughly isometric if the metric spaces V_1 and V_2, with the usual geodesic distances, are roughly isometric.

§2. Some lemmas

In the following we will be concerned with graphs of bounded degree i.e., graphs such that

$$(7.4) \qquad\qquad \sup_{x \in V} \deg(x) < M \quad \text{for some constant } M.$$

In the paper by Markvorsen, Mc Guiness and Thomassen cited above it was proved, among other things, the discrete counterpart of Kanai's result: if Γ_1 and Γ_2 are roughly isometric graphs of bounded degree, then Γ_2 is recurrent if Γ_1 is. On the other hand it was proved in Soardi (1993)(a) that rough isometries preserve another important property: if Γ_1 and Γ_2 are roughly isometric graphs of bounded degree, then $\Gamma_1 \in \mathcal{O}_{HD}$ implies $\Gamma_2 \in \mathcal{O}_{HD}$.

In this section we prepare ourselves for the proof of these results. Since we shall be dealing with Dirichlet spaces on several graphs Γ, we will denote by $\mathbf{D}(\Gamma)$, $\mathbf{HD}(\Gamma)$ etc. such spaces. Analogously, we will denote by d_Γ the distance on the vertex set of Γ.

Let us start with the definition of the k–fuzz of a graph (see Doyle and Snell (1984)).

DEFINITION. Let $\Gamma = (V, Y)$ be a locally finite graph and let k be any positive integer. The k–fuzz of Γ is the graph $\Gamma^k = (V^k, Y^k)$ such that $V^k = V$ and

$$Y^k = \{[x, y] \in V \times V : d_\Gamma(x, y) \leq k\}.$$

In other words we join by a new edge every couple of vertices of Γ whose distance is not larger than k (in particular we assign a self–loop to every vertex). The introduction of the k–fuzz of a graph is motivated by the following observation. Suppose that Γ_1 and Γ_2 are roughly isometric graphs and let $\phi : V_1 \mapsto V_2$ be the rough isometry. Then it is easily seen by (7.1) that ϕ is a morphism from Γ_1 to Γ_2^k for all $k \geq a + b$. Since

$$d_{\Gamma^k}(x, y) \leq d_\Gamma(x, y) \leq k d_{\Gamma^k}(x, y) \quad \text{for all } x, y \in V,$$

ϕ is still a rough isometry from Γ_1 to Γ_2^k.

Lemma (7.5). *Suppose that Γ is of bounded degree and let k be any positive integer. Then Γ is recurrent if and only if Γ^k is recurrent. Moreover $\Gamma \in \mathcal{O}_{\mathbf{HD}}$ if and only if $\Gamma^k \in \mathcal{O}_{\mathbf{HD}}$.*

Proof. First we prove that the norms in $\mathbf{D}(\Gamma)$ and in $\mathbf{D}(\Gamma^k)$ are equivalent.

On one hand it is clear that, for every function f on $V = V^k$, $\|f\|_{\mathbf{D}(\Gamma)} \leq \|f\|_{\mathbf{D}(\Gamma^k)}$. On the other hand, set for every edge $e = [z_1, z_2] \in Y^k$

$$U(e, k) = \{[x_1, x_2] \in Y : d(x_j, z_i) \leq k, \text{ for } i, j = 1, 2\}.$$

Clearly, every edge in Γ belongs to less than $M^{2(k+1)}$ sets $U(e, k)$, where M is as in (7.4). Now, if $1 \leq d(x, t) = n \leq k$, then there is a path in Γ with vertices $x_0 = x \sim x_1 \sim \cdots \sim x_n = t$. Then we have

$$| f(x) - f(t) |^2 \leq k \sum_{j=1}^{n} | f(x_j) - f(x_{j-1}) |^2$$

$$\leq k \sum_{[s, u] \in U([x, t], k)} | f(s) - f(u) |^2 .$$

It follows

$$\sum_{[x,t]\in Y^k} |\, f(x) - f(t)\,|^2 \le k \sum_{[x,t]\in Y^k} \sum_{[s,u]\in U([x,t],k)} |\, f(s) - f(u)\,|^2$$

$$\le kM^{2(k+1)} \sum_{[s,u]\in Y} |\, f(s) - f(u)\,|^2 \,.$$

Hence

$$\|\, f\,\|_{\mathbf{D}(\Gamma^k)} \le k^{1/2} M^{(k+1)} \|\, f\,\|_{\mathbf{D}(\Gamma)} \,.$$

By Theorem (3.63) we may conclude that Γ is recurrent if and only if Γ^k is.

Now assume that Γ^k is transient and belongs to $\mathcal{O}_{\mathbf{HD}}$. Let $h \in \mathbf{HD}(\Gamma)$. By Royden's decomposition theorem (Theorem 3.69) there is $v \in D_0(\Gamma^k)$ and a constant κ such that $h = v + \kappa$. Since v belongs also to $D_0(\Gamma)$ and the Royden decomposition is unique, $v = 0$ and $h = \kappa$. Hence $\Gamma \in \mathcal{O}_{\mathbf{HD}}$. The same argument proves also the converse implication. $\square$

DEFINITION. Let ψ be a morphism from Γ_1 to Γ_2. The image $\Gamma'_1 = (V'_1, Y'_1)$ of Γ_1 in Γ_2 by means of ψ is the graph such that $V'_1 = \psi(V_1)$, and $[x', t'] \in Y'_1$ if there exists $[x, t] \in Y_1$ such that $\psi(x) = x'$ and $\psi(t) = t'$.

It is clear that Γ'_1 is a connected subgraph of Γ_2.

Lemma (7.6). *Let Γ and Γ' be graphs of bounded degree. Suppose that there is a morphism ψ of Γ onto Γ' in such a way that Γ' coincides with the image of Γ by means of ψ. Suppose moreover that there is a constant m such that, for every $x, y \in V$, $\psi(x) = \psi(y)$ implies $d_\Gamma(x, y) \le m$. Under these assumptions, if Γ is transient, then Γ' is transient as well. If $\Gamma \in \mathcal{O}_{\mathbf{HD}}$, then $\Gamma' \in \mathcal{O}_{\mathbf{HD}}$ as well.*

Proof. Set $\Gamma = (V, Y)$ and $\Gamma' = (V', Y')$. For every function f' on V' we define a function f on V by setting

$$(7.7) \qquad f(x) = f'(\psi(x)), \qquad \text{for all } x \in V.$$

Let $\mathbf{C}$ denote the space of all Dirichlet finite functions on V of the form (7.7). We will show that (7.7) defines a Banach space isomorphism of $\mathbf{C}$ onto $\mathbf{D}(\Gamma')$.

We have

$$(7.8) \quad \sum_{x \sim y} |f(x) - f(y)|^2 \le \sum_{\substack{x' \sim y' \\ x', y' \in V(G')}} \sum_{\substack{\psi(x) = x' \\ \psi(y) = y'}} |f(x) - f(y)|^2$$

$$= \sum_{\substack{x' \sim y' \\ x', y' \in V(G')}} |f'(x') - f'(y')|^2 \sum_{\substack{x \in \psi^{-1}(x') \\ y \in \psi^{-1}(y')}} 1.$$

Let $s' \in V'$ and let $s \in V$ be such that $\psi(s) = s'$. Then $\psi^{-1}(s') \subseteq \{x \in V : d_\Gamma(x, s) \leq m\}$. Therefore, denoting by $|\cdot|$ the cardinality, $|\psi^{-1}(s')| \leq M^{m+1}$ and by (7.8) we get (choosing $o' = \psi(o)$)

$$(7.9) \qquad D(f) \leq M^{2(m+1)} D(f') \qquad \text{whence} \qquad \|f\|_{\mathbf{D}(\Gamma)} \leq M^{m+1} \|f'\|_{\mathbf{D}(\Gamma')}.$$

Since Γ' coincides with the image of Γ, for every $x', y' \in V'$, $x' \sim y'$, there exist $x, y \in V$ such that $x \sim y$ and $\psi(x) = x'$, $\psi(y) = y'$. It follows that

$$(7.10) \qquad\qquad D(f') \leq D(f) \qquad \text{whence} \qquad \|f'\|_{\mathbf{D}(\Gamma')} \leq \|f\|_{\mathbf{D}(\Gamma)}$$

Suppose now that Γ' is recurrent. Then, by (2) of Theorem (3.63) there is a sequence of finitely supported functions f'_n on V' such that

$$(7.11) \qquad\qquad \|e' - f'_n\|_{\mathbf{D}(\Gamma')} \to 0 \quad \text{as } n \to \infty,$$

where e' is the function identically equal to 1 on V'. Let f_n and f'_n be related as in (7.7). Then f_n is finitely supported. By (7.9), (7.11) and (2) of Theorem (3.63) again, we see that Γ is recurrent.

Now we claim that if f and f' are as in (7.7), then $f \in \mathbf{C} \bigcap \mathbf{D}_0(G)$ if and only if $f' \in \mathbf{D}_0(G')$.

One implication is obvious by (7.9). To prove the converse, write V as the disjoint union of the classes $\psi^{-1}(x')$ with $x' \in V'$. For every $x \in V$ let $\{x\}$ be the class of x and let $\omega(x)$ be the cardinality of $\{x\}$. For every $g \in \mathbf{D}(\Gamma)$ we define the mean $\mathcal{M}g$ of g by

$$\mathcal{M}g(x) = \frac{1}{\omega(x)} \sum_{s \in \{x\}} g(s), \qquad \text{for all } x \in V.$$

Suppose that $x \sim y$. Then

$$(7.12) \qquad \begin{aligned} |\mathcal{M}g(x) - \mathcal{M}g(y)|^2 &= (\omega(x)\omega(y))^{-2} \Big(\sum_{\substack{s \in \{x\} \\ t \in \{y\}}} g(s) - g(t) \Big)^2 \\ &\leq (\omega(x)\omega(y))^{-1} \sum_{\substack{s \in \{x\} \\ t \in \{y\}}} |g(s) - g(t)|^2. \end{aligned}$$

Clearly, $d_\Gamma(s, t) \leq 2m + 1$ for every $s \in \{x\}$ and $t \in \{y\}$. Hence, there exists a path in Γ with vertices $s = x_0 \sim x_1 \sim \cdots \sim x_n = t$ such that $n \leq 2m + 1$. It follows that

$$(7.13) \qquad |g(s) - g(t)|^2 \leq (2m + 1) \sum_{j=1}^{n} |g(x_j) - g(x_{j-1})|^2.$$

We set, for every couple of vertices z_1, z_2 in V such that $d_\Gamma(z_1, z_2) \leq 2m + 1$,

$$N(z_1, z_2) = \{[x_1, x_2] \in Y : d_\Gamma(x_j, z_i) \leq 2m + 1, \quad \text{for } i, j = 1, 2\}.$$

Every edge $[x_1, x_2] \in Y$ belongs to less than $M^{4(m+1)}$ sets $N(z_1, z_2)$ so that by (7.13) above,

$$|g(s) - g(t)|^2 \leq (2m + 1) \sum_{[x_1, x_2] \in N(s,t)} |g(x_1) - g(x_2)|^2.$$

Taking (7.12) into account,

$$\sum_{x \sim y} |\mathcal{M}g(x) - \mathcal{M}g(y)|^2 \leq (2m + 1) \sum_{x \sim y} \sum_{\substack{s \in \{x\} \\ t \in \{y\}}} \sum_{[x_1, x_2] \in N(s,t)} |g(x_1) - g(x_2)|^2$$

$$\leq (2m + 1) M^{4(m+1)} \sum_{[x_1, x_2] \in Y} |g(x_1) - g(x_2)|^2.$$

Hence,

(7.14) $$D(\mathcal{M}g) \leq (2m + 1) M^{4m+4} D(g).$$

Moreover, it is easy to see that

(7.15) $$|g(s)|^2 \leq (m + 1)(D(g) + |g(o)|^2), \quad \text{for every } s \in \{o\}.$$

Hence,

(7.16) $$|\mathcal{M}g(o)|^2 \leq \frac{1}{\omega(o)} \sum_{s \in \{o\}} |g(s)|^2 \leq (m + 1)\|g\|^2_{\mathbf{D}(\Gamma)}.$$

Finally, from (7.14) and (7.16) we get

(7.17) $$\|\mathcal{M}g\|_{\mathbf{D}(\Gamma)} \leq \sqrt{3m + 2} M^{2m+2} \|g\|_{\mathbf{D}(\Gamma)}.$$

Assume that f is a function in $\mathbf{C} \cap \mathbf{D}_0(\Gamma)$. Let $f_n \in \ell_0(\Gamma)$ be a sequence such that $\|f - f_n\|_{\mathbf{D}(\Gamma)} \to 0$. Since $\mathcal{M}f = f$, (7.17) implies $\|f - \mathcal{M}f_n\|_{\mathbf{D}(\Gamma)} \to 0$. Therefore, our claim follows from (7.10).

Since, as already proved, the recurrence of Γ' implies that of Γ, we may assume that Γ' is transient. Suppose $f' \in \mathbf{HD}(\Gamma')$. By Theorem (3.69) there is a constant κ and a function $g \in \mathbf{D}_0(\Gamma)$ such that $f = g + \kappa$. If g and g' are related as f and f' in (7.7), we have that $g' \in \mathbf{D}_0(\Gamma')$ and $f' = g' + \kappa$. By the uniqueness of Royden's decomposition in $\mathbf{D}(\Gamma')$, $g' = 0$ and $f' = \kappa$. Hence, $\Gamma' \in \mathcal{O}_{\mathbf{HD}}$ □

REMARK. Note that we can not use Theorem (3.45) to prove that the transience of Γ implies the transience of Γ'. Actually, all the resistances are equal to 1 in both graphs so that ψ cannot be a network morphism (unless it is an isomorphism); see §1 in Chapter I.

§3. Recurrence, class $\mathcal{O}_{HD}$ and rough isometries

We start with the first result announced in §2.

Theorem (7.18). *Let Γ_1 and Γ_2 be roughly isometric graphs of bounded degree. Then Γ_2 is recurrent if so is Γ_1.*

Proof. Let ϕ be a rough isometry from Γ_1 to Γ_2. Let k be positive integer such that $k > a + b$, where a and b are as in (7.1). Then ϕ is a rough isometry and a morphism from Γ_1 to the k–fuzz Γ_2^k of Γ_2. Let Γ_1' be the image of Γ_1 in Γ_2 by means of ϕ. For every couple of vertices x and y of Γ_1 such that $\phi(x) = \phi(y)$ we have by (7.1)

$$a^{-1} d_{\Gamma_1'}(x,y) - b \le d_{\Gamma_2}(\phi(x), \phi(y)) = 0,$$

so that $d_{\Gamma_1'}(x,y) \le ab$. Thus we may apply Lemma (7.6) with $m = ab$. Therefore Γ_1' is transient if so is Γ_1. It follows from (2) of Corollary (3.48) that Γ_2^k is transient. By Lemma (7.5) Γ_2 is transient. $\square$

EXAMPLE (7.19). Let Γ be a plane graph. Assume that Γ is uniformly embedded in $\mathbb{R}^2$ i.e., there is a constant $\kappa > 0$ such that, denoting by $\|\cdot\|$ the euclidean norm,

$$\kappa^{-1}\|z_1 - z_2\| \le d(z_1, z_2) \le \kappa\|z_1 - z_2\| \quad \text{for all } z_1, z_2 \in V$$

(compare with (4.40)). Then Γ is roughly isometric to $\mathbb{R}^2$. Since Z^2 is uniformly embedded in $\mathbb{R}^2$ as well, by Theorem (7.18) Γ is recurrent.

The graphs studied in §7 of Chapter V, even though transient, in general are not uniformly embedded in the plane: see e.g. Example (5.70).

The proof that any rough isometry transforms the class $\mathcal{O}_{HD}$ into itself is more laborious.

Lemma (7.20). *Let $\Gamma = (V, Y)$ be a transient subgraph of a graph $\Lambda = (L, E)$ satisfying (7.4). Assume that the inclusion map of Γ into Λ is a rough isometry. If $\Gamma \in \mathcal{O}_{HD}$, then $\Lambda \in \mathcal{O}_{HD}$.*

Proof. Let $f \in \mathbf{HD}(\Lambda)$. We will show that there is a constant γ such that, for almost all paths $\mathbf{p} \in \mathbf{P}_o$, $f(x)$ tends to γ as $x \to \infty$ along the vertices of $\mathbf{p}$. By Corollary (3.95) this will imply that f is a constant, thus proving the thesis of the Lemma.

Let g denote the restriction of f to V. Clearly, $g \in \mathbf{D}(\Gamma)$. Let $\mathbf{P}_o^\Gamma$ denote the set of all infinite paths in $\mathbf{P}_o$, all of whose edges are in Y (we may assume $o \in V$). Denote, for every Dirichlet finite function h, by $h(\mathbf{p})$ the limit of $h(x)$ as x tends to infinity along the vertices of the one–ended infinite path $\mathbf{p}$ (see Theorem (3.85)).

Since $\Gamma \in \mathcal{O}_{HD}$, by Royden's decomposition theorem and Theorem (3.86) there exists a constant γ such that $g(\mathbf{p}) = \gamma$, for almost every path $\mathbf{p}$ in $\mathbf{P}_o^\Gamma$. We set

$$\mathbf{P}^I = \{\mathbf{p} \in \mathbf{P}_o : f(\mathbf{p}) = \gamma\}$$
$$\mathbf{P}^{II} = \{\mathbf{p} \in \mathbf{P}_o : f(\mathbf{p}) \neq \gamma\}$$
$$\mathbf{P}^{III} = \{\mathbf{p} \in \mathbf{P}_o : f(\mathbf{p}) \text{ does not exist}\}$$
$$\mathbf{Q} = \mathbf{P}^{II} \cap \mathbf{P}_o^\Gamma.$$

Therefore almost every path $\mathbf{p} \in \mathbf{P}_o^\Gamma$ belongs to $\mathbf{P}^I$. Moreover $\lambda_\Lambda(\mathbf{P}^{III}) = \infty$ by Theorem (3.85). To prove the Lemma we have only to show that $\lambda_\Lambda(\mathbf{P}^{II}) = \infty$ (note that Λ is transient so that, by Corollary (3.84), $\lambda_\Lambda(\mathbf{P}_o) < \infty$).

In what follows, we will write $x \sim y$ if $d_\Lambda(x,y) = 1$ and $x \approx y$ if $d_\Gamma(x,y) = 1$. Now, we associate to every $\mathbf{p} \in \mathbf{P}^{II}$ a path $\tilde{\mathbf{p}}$ in $\mathbf{Q}$. Choose any path $\mathbf{p} \in \mathbf{P}^{II}$ with distinct vertices $o = x_0 \sim x_1 \sim \cdots x_n \sim \cdots$. By (7.2), for every n there exist vertices $y_n \in V$ (not necessarily distinct) such that $d_\Lambda(x_n, y_n) \leq c$. Let $\rho = a(2c + 1 + b)$, where a, b, c are as in (7.1), (7.2). We have by (1)

$$\begin{aligned} d_\Gamma(y_n, y_{n+1}) &\leq a d_\Lambda(y_n, y_{n+1}) + ab \\ (7.21) \qquad &\leq a\big(d_\Lambda(y_n, x_n) + d_\Lambda(x_n, x_{n+1}) + d_\Lambda(x_{n+1}, y_{n+1})\big) + ab \\ &\leq a(2c + 1 + b) = \rho. \end{aligned}$$

Note also that $d_\Lambda(y_n, y_{n+1}) \leq d_\Gamma(y_n, y_{n+1})$.

Suppose $y_n \neq y_{n+1}$. By (7.21), there are distinct vertices $z_r \in V$, $r = 0, \ldots, k$, with $k \leq \rho$, such that

$$y_n = z_0 \approx z_1 \cdots \approx z_k = y_{n+1}.$$

It follows that there exists a path $\tilde{\mathbf{p}} \in \mathbf{P}_o^\Gamma$ with distinct vertices $t_j \in V$,

$$o = t_0 \approx t_1 \approx \cdots t_j \approx \cdots$$

and a nondecreasing sequence of subscripts $j(n) \to \infty$ such that

$$(7.22) \qquad t_{j(n)} = y_n, \qquad j(n+1) - j(n) \leq \rho, \qquad d_\Lambda(x_n, t_{j(n)}) \leq c.$$

Note that, for every integer r, the cardinality of the set $\{n : j(n) = r\}$ is smaller than M^{c+1}.

For every n there are vertices $v_r \in L$, $r = 0, \ldots, m$, with $m \leq c$, such that

$$v_0 = x_n \sim v_1 \sim \cdots v_m = t_{j(n)}.$$

Then, as $n \to \infty$,

$$|f(x_n) - f(t_{j(n)})|^2 \le c \sum_{r=1}^{m} |f(v_r) - f(v_{r-1})|^2 \to 0$$

so that $g(t_{j(n)}) = f(t_{j(n)}) \to f(\mathbf{p}) \ne \gamma$. In the same way, we see that the whole sequence $g(t_j)$ tends to $f(\mathbf{p})$ since, for $j(n) < j < j(n+1)$, we have

$$|f(t_j) - f(t_{j(n)})|^2 \le \rho \sum_{r=1}^{k} |f(z_r) - f(z_{r-1})|^2 \to 0.$$

It follows that the path $\bar{\mathbf{p}}$ with vertices t_j belongs to $\mathbf{Q}$.

Since $\lambda_\Gamma(\mathbf{Q}) = \infty$, by (3) of Lemma (3.75), there exists a positive function w on Y such that

(7.23)
$$\sum_{\tilde{e} \in Y} w^2(\tilde{e}) = \mathcal{E}(w) < \infty,$$

(7.24)
$$\sum_{\tilde{e} \in Y(\mathbf{p})} w(\tilde{e}) = \infty, \qquad \text{for all } \mathbf{p} \in \mathbf{Q}$$

where $\mathbf{p}$ has vertex and edge sets $V(\mathbf{p})$ and $Y(\mathbf{p})$. Let ρ be as above and set $\delta = \rho + c$. For every edge $e = [z_1, z_2] \in E$ set

$$U(e, \delta) = \{\tilde{e} = [x_1, x_2] \in Y : d_\Lambda(x_j, z_i) \le \delta, \quad \text{for } i, j = 1, 2\}.$$

We define w^* on E in the following way:

$$w^*(e) = \sup_{\tilde{e} \in U(e, \delta)} w(\tilde{e}), \qquad \text{for all } e \in E.$$

Clearly, $w^{*2}(e) \le \sum_{\tilde{e} \in U(e, \delta)} w^2(\tilde{e})$ for all $e \in E$. Moreover, every edge $\tilde{e} \in Y$ belongs to less than $M^{2(\delta+1)}$ sets $U(e, \delta)$, $e \in E$. Therefore, (7.23) gives

$$\mathcal{W}(w^*) = \sum_{e \in E} w^{*2}(e) \le \sum_{e \in E} \sum_{\tilde{e} \in U(e, \delta)} w^2(\tilde{e})$$

(7.25)
$$\le M^{2(\delta+1)} \sum_{\tilde{e} \in Y} w^2(\tilde{e})$$

$$< \infty.$$

We now claim that

(7.26)
$$\sum_{e \in Y(\mathbf{p})} w^*(e) = \infty, \qquad \text{for all } \mathbf{p} \in \mathbf{P}^{II}.$$

Let, for every $\mathbf{p} \in \mathbf{P}^{II}$ with vertices x_n, $\bar{\mathbf{p}}$ be the path in $\mathbf{Q}$ with vertices t_j constructed above. For such a $\bar{\mathbf{p}}$, $\sum_{\bar{e} \in Y(\bar{\mathbf{p}})} w(\bar{e}) = \infty$, by (21). For every edge $[x_n, x_{n+1}] \in Y(\mathbf{p})$, such that $j(n) \neq j(n+1)$,

$$(7.27) \qquad \rho w^*([x_n, x_{n+1}]) \geq \sum_{j=j(n)+1}^{j(n+1)} w([t_{j-1}, t_j])$$

since, by (7.21) and (7.22), $[t_{j-1}, t_j] \in U([x_n, x_{n+1}], \delta)$ for $j = j(n)+1, \ldots, j(n+1)$. It follows, from (7.24) and (7.27), that

$$\sum_{e \in Y(\mathbf{p})} w^*(e) \geq \rho^{-1} \sum_{\bar{e} \in Y(\bar{\mathbf{p}})} w(\bar{e}) = \infty.$$

Therefore, by (7.25), (7.26) and (3) of Lemma (3.75) again, $\lambda_\Lambda(P^{II}) = \infty$. Hence, $f(\mathbf{p}) = \gamma$ for almost every path in $\mathbf{P}_o$. $\square$

Theorem (7.28). *Suppose that Γ_1 and Γ_2 are roughly isometric graphs of bounded degree. If $\Gamma_1 \in \mathcal{O}_{HD}$, then $\Gamma_2 \in \mathcal{O}_{HD}$.*

Proof. We may assume that both the graphs are transient. Let ϕ be a rough isometry from Γ_1 to Γ_2. As in the proof of Theorem (7.18) ϕ is a morphism from Γ_1 to the k-fuzz Γ_2^k of Γ_2 and we may apply Lemma (7.6). It follows that Γ_2^k contains a roughly isometric subgraph belonging to $\mathcal{O}_{HD}$. Therefore, by Lemma (7.20), $\Gamma_2^k \in \mathcal{O}_{HD}$. Finally, by Lemma (7.5), $\Gamma_2 \in \mathcal{O}_{HD}$. $\square$

REMARK. The continuous version of Theorem (7.28) is due to Holopainen (1994) who proved that if M and N are roughly isometric complete Riemann manifolds of bounded geometry and if $M \in \mathcal{O}_{HD}$, then $N \in \mathcal{O}_{HD}$ as well. Holopainen's remarkable result is in fact much more general, because it deals not only with harmonic functions but also with p–harmonic functions (see §2 in the Appendix). For an earlier result in this direction see Pansu (1989).

We notice that Markvorsen, Mc Guinness and Thomassen (1992) proved the conclusion of Theorem (7.28) under the extra assumption that the graphs Γ_j satisfy a dimensional isoperimetric inequality (j=1,2).

As an application of Theorem (7.28), we will now improve Theorem (4.75) (see Medolla and Soardi (1993)).

Theorem (7.29). *Let (Γ, r) be a network such that $r(x, y) = 1$ for all $x \sim y$. Assume that Γ has moderate growth and that $\mathrm{Aut}(\Gamma)$ acts on V with a finite number of orbits. Then $\Gamma \in \mathcal{O}_{HD}$.*

Proof. Let $K_1, \ldots, K_n$ denote the orbits. Let

$$m = \max_{i,k} \min\{d(x, y) : x \in K_i, y \in K_k\}.$$

For any $h > 2m + 1$, let Γ^h be the h–fuzz of Γ. For every $x \in K_i$ and every k there is $y \in K_k$ such that x and y are neighbors in Γ^h, $i, k = 1, \ldots, n$. Furthemore, as remarked in Salvatori (1992), K_i induces a connected vertex transitive subgraph Υ_i of Γ^h.

It is easily seen that Γ^h has the same properties as Γ. In fact, $\mathrm{Aut}(\Gamma^h)$ acts with a finite number of orbits. Moreover, by Lemma 2 in Salvatori's paper (or by Theorem (7.34) below) Γ^h satisfies a strong isoperimetric inequality if and only if Γ does. Hence Γ^h has moderate growth by Theorem (4.39).

We claim that Υ_i has moderate growth as well. Namely, let $\{U_k\}$ be a sequence of finite sets in V such that $U_k \subseteq U_{k+1}$, $\bigcup_k U_k = V$ and

$$(7.30) \qquad \lim_{k \to \infty} \frac{|\daleth^h(U_k)|}{|U_k|} = 0,$$

where $\daleth^h(U_k)$ is the combinatorial boundary of U_k in Γ^h. Set $U_{k,i} = U_k \cap K_i$. Then the sequence $\{U_{k,i}\}$ is increasing to K_i. We have also the analogue of (7.30). To see this denote by $V^h(y)$, $y \in V$, the set of all neighbors of y in Γ^h. Denote also by $\daleth_i U_{k,i}$ the combinatorial boundary of $U_{k,i}$ in Υ_i. Then

$$U_k \subseteq \bigcup_{y \in U_{k,i}} V^{(h)}(y)$$

$$\daleth_i U_{k,i} \subseteq \daleth^h U_k,$$

Therefore $|\daleth_i U_{k,i}| \le |\daleth^h U_k|$ and $|U_k| \le M |U_{k,i}|$, where M is the maximum degree in Γ^h. Hence Υ_i has moderate growth by (7.30). By Theorem (4.65) $\Upsilon_i \in \mathcal{O}_{\mathbf{HD}}$.

Let $d(x,y)$, $d^{(h)}(x,y)$, and $d_1(x,y)$ denote the distances in Γ, Γ^h and in Υ_1, respectively. Let $x, y \in K_1$ be such that $d(x,y) = r$. Let $x = x_0 \sim x_1 \sim \ldots \sim x_r = y$ be the vertices of a path in Γ joining x and y.

There are vertices $y_j \in K_1$ such that $d(y_j, x_j) \le m$ for all $j = 1, \ldots, r-1$. Then $d(y_j, y_{j+1}) \le 2m + 1$, so that y_j and y_{j+1} are neighbors in Υ_1. Hence $d_1(x,y) \le r$. On the other hand $d^{(h)}(x,y) \ge h^{-1} r$. It follows

$$(7.31) \qquad d^{(h)}(x,y) \le d_1(x,y) \le h d^{(h)}(x,y) \quad \text{for all } x, y \in X_1.$$

Hence the inclusion map is a rough isometry of Υ_1 into Γ^h. The thesis now follows from Lemma (7.5) and Theorem (7.28).

§4. Combinatorial boundaries

In this section we show that rough isometries preserve strong isoperimetric inequalities and moderate growth; compare with Kanai (1986)(b). We start with a property already noticed by Salvatori (1992).

Lemma (7.32). *Let Γ be an infinite graph of bounded degree. Let $\daleth(L)$ and $\daleth^k(L)$ denote the combinatorial boundaries in Γ and in Γ^k, respectively, of a subset of vertices L. Then, there exists a positive constant $\kappa_1 > 0$ such that*

$$(7.33) \qquad |\daleth(L)| \leq |\daleth^k(L)| \leq \kappa_1 |\daleth(L)|.$$

Proof. It is clear that, for every finite subset L, $\daleth(L) \subseteq \daleth^k(L)$. Hence the first inequality in (7.33) is obvious. On the other hand, for any finite subset L, $\daleth^k(L)$ consists of all the vertices of L at distance less than or equal to $k - 1$ from $\daleth(L)$. If M is as in (7.4), then

$$|\daleth^k(L)| \leq |\daleth(L)| \sum_{j=1}^{k-1} M^j.$$

$\square$

Theorem (7.34). *Let Γ_1 and Γ_2 be roughly isometric graphs of bounded degree. If Γ_1 satisfies a strong isoperimetric inequality, then Γ_2 satisfies a strong isoperimetric inequality as well.*

Proof. Let ϕ be a rough isometry from Γ_1 to Γ_2. If $k > \max(a + b, c)$, then ϕ is also a morphism of Γ_1 into Γ_2^k. By Lemma (7.32) Γ_2 satisfies a strong isoperimetric inequality if and only if Γ_2^k does. So, we may assume that ϕ is also a morphism of Γ_1 into Γ_2 and that the constant c in (7.2) is 1.

Let $\Gamma' \subseteq \Gamma_2$ be the image of Γ_1 by means of ϕ. If $m > ab$ we have from (7.1)

$$(7.35) \qquad \phi(x) = \phi(y) \text{ implies } d_{\Gamma_1}(x, y) \leq m.$$

Let V_1 and V' denote the vertex sets of Γ_1 and Γ', respectively. Let f' be a finitely supported function on V' and set $f(x) = f'(\phi(x))$, $x \in V_1$. By (7.35) f is finitely supported on V. Denote by M the maximum degree in Γ_1. Then, by Theorem (4.27) ($(1) \Leftrightarrow (2)$) and by (7.9)

$$\sum_{x \in V'} c(x) f'^2(x) \leq M \sum_{y \in V_1} c(y) f^2(y)$$
$$\leq \gamma M D(f) \leq \gamma M^{2m+3} D(f').$$

Therefore, always by Theorem (4.27), Γ' satisfies a strong isoperimetric inequality.

Denote by V_2 the vertex set of Γ_2. Let f be a finitely supported function on V_2 and denote by g the restriction of f to V'. For every $x \in V_2 \setminus V'$ there is $y \in V'$ such that $x \sim y$. Hence $|f(x)|^2 \leq 2|g(y)|^2 + 2|f(x) - f(y)|^2$. We then have

$$(7.36) \qquad \sum_{x \in V_2} c(x) |f(x)|^2 \leq 2M^2 \sum_{x \in V'} c(y) |g(y)|^2 + 2D(f)$$
$$\leq 2M^2 \gamma D(g) + 2D(f),$$

where $D(g)$ is the Dirichlet sum of g on Γ'. As $D(g) \leq D(f)$, the thesis follows from (7.36) and Theorem (4.27). $\square$

Theorem (7.37). *Let Γ_1 and Γ_2 be roughly isometric graphs of bounded degree. If Γ_2 has moderate growth, then Γ_1 has moderate growth too.*

Proof. Let ϕ be a rough isometry from Γ_1 to Γ_2. Taking Lemma (7.32) into account, we may assume, as in the proof of Theorem (7.34), that ϕ is a morphism from Γ_1 to Γ_2 and that the constant c in (7.2) is 1. Moreover ϕ satisfies (7.35).

Let Γ' be defined as in the proof of Theorem (7.34). We first prove that if Γ_2 has moderate growth, then the same holds for Γ'. Suppose that $\{U_n\}$ is an increasing sequence of subsets of V_2 such that $\cup_n U_n = V_2$ and such that $|\daleth_2(U_n)|/|U_n| \to 0$ as $n \to \infty$ (where $\daleth_2(U_n)$ is the boundary of U_n in V_2).

Set $U'_n = U_n \cap V'$. Then $U'_n \subseteq U'_{n+1}$ and $\cup_n U'_n = V'$. Denoting by $\daleth'(U'_n)$ the combinatorial boundary of U'_n in V', we have

$$(7.38) \qquad\qquad |\daleth'(U'_n)| \leq |\daleth_2(U_n)| \quad \text{for all } n.$$

Now, for every $x \in U_n \setminus \daleth_2(U_n)$ there is $y \in U'_n$ such that $x \sim y$. It follows

$$|U_n \setminus \daleth_2(U_n)| \leq M|U'_n|,$$

where M is as in the maximum degree in Γ_2. Combining this with (7.38) we have that $|\daleth'(U'_n)|/|U'_n| \to 0$ as $n \to \infty$. Thus Γ' has moderate growth.

Finally we prove that if Γ' has moderate growth, then also Γ_1 has moderate growth. Let U'_n be as above and set $A_n = \phi^{-1}(U'_n)$. Then A_n is an increasing sequence whose union is V_1.

Suppose $y \in U'_n$, $x \notin U'_n$ and $x \sim y$. Let $z \in A_n$ be such that $\phi(z) = y$. Since ϕ is a morphism and a surjection of Γ_1 onto Γ', there must exists $s \notin A_n$, $s \sim z$ such that $\phi(s) = x$. Therefore $\phi(\daleth_1(A_n)) \supseteq \daleth'U'_n$, where $\daleth_1(A_n)$ denotes the boundary of A_n.

On the other hand, suppose that there is $z \in \daleth_1(A_n)$ such that $\phi(z) \notin \daleth'U'_n$. Then there is $s \sim z$, $s \notin A_n$, such that $\phi(s) \in U'_n$, contradicting the definition of A_n. Therefore

$$(7.39) \qquad\qquad \phi(\daleth_1(A_n)) = \daleth'(U'_n).$$

Since $|\phi^{-1}(x)| \leq M^{m+1}$ for all $x \in V'$, (7.39) implies $|\daleth_1(A_n)| \leq M^{m+1}|\daleth'(U'_n)|$. As $|A_n| \geq |U'_n|$, Γ_1 has moderate growth. $\square$

APPENDIX

NONLINEAR NETWORKS

1. Nonlinear networks and modular sequence spaces

The theory studied in the preceding chapters has a natural extension to nonlinear networks. In this appendix we will outline, without proofs, a theory of such networks. The first section is based on De Michele and Soardi (1990), Soardi (1993)(b) and Zemanian (1991).

A nonlinear resistive network is a network such that the voltage across the endpoints of an edge B is a nonlinear function $p_B(i(B))$ (the resistance function) of the current $i(B)$ flowing in the branch. The mathematical setting is provided by the modular sequence spaces studied by Woo (1973). Such spaces generalize the Orlicz sequence spaces, which in turn generalize the ℓ^p spaces.

More precisely, a nonlinear network is defined in the following way. Let $\Gamma = (V, Y)$ be a connected graph, which, for simplicity we suppose locally finite and without self–loops (see Affer (1992), Soardi (1993)(b) and Soardi and Yamasaki (1993) for the general case).

Suppose that every edge B of Γ is assigned a continuous function $p_B : \mathbb{R} \mapsto \mathbb{R}$ which is odd, strictly increasing, and such that $\lim_{i \to \pm\infty} p_B(i) = \pm\infty$.

Denote by $\mathcal{P}$ the set of such functions and assume also that $p_{[x,y]} = p_{[y,x]}$ for all $x \sim y$.

DEFINITION. The pair $(\Gamma, \mathcal{P})$ will be called nonlinear network with resistance functions p_B.

Clearly, if $p_B(i) = r_B i$, then the network reduces to a linear network as defined in §1 of chapter I.

We will assume that the following uniform variation condition (a generalization of the Δ_2 condition for Orlicz spaces) is satisfied by the family $\mathcal{P}$: there exist a number r and a finite set $F \subset Y$ such that

(A.1) $$\frac{t p_B(t)}{M_B(t)} \leq r \qquad \text{for all} \quad t \in (0, a_B) \quad \text{and} \quad B \notin F$$

where

(A.2) $$M_B(t) = \int_0^t p_B(u) \, du$$

and a_B is the unique positive number such that $M_B(a_B) = 1$.

Observe that, denoting by q_B the inverse function of p_B and setting

$$M_B^*(t) = \int_0^t q_B(u)\, du,$$

M_B and M_B^* are complementary Orlicz functions.

In order to formulate Kirchhoff's laws we summarize the relevant parts of the theory of modular sequence spaces.

We set for every 1–chain I

(A.3) $$\mathcal{W}(I) = \frac{1}{2} \sum_{B \in Y} M_B(i(B)),$$

where M_B is as in (A.2). The modular sequence space $\ell\{M_B\}$ is defined as the linear space of all 1–chains I such that, for some $t > 0$,

$$\sum_{B \in Y} M_B\left(\frac{i(B)}{t}\right) < \infty.$$

For all $I \in \ell\{M_B\}$ set

$$\|I\| = \frac{1}{2} \inf\{t : \sum_{B \in Y} M_B\left(\frac{i(B)}{t}\right) \le 1\}.$$

With this norm $\ell\{M_B\}$ becomes a Banach space. Note that this definition does not require condition (A.1). However, if (A.1) holds, then it is possible to show that $\mathcal{W}(I)$ is finite for every $I \in \ell\{M_B\}$ and that the elementary 1–chains $\{\chi(B)\}$, $B \in X$ (see §2 in chapter I), form an unconditional basis for the modular sequence space.

In the linear case $\ell\{M_B\} = \ell^2$ and the functional $\mathcal{W}(I)$ in (A.3) coincides with the usual energy of a 1–chain (except for an inessential factor $1/2$). In the general case the space $\mathbf{H}_1$ is replaced by the space $\ell\{M_B\}$ and the functional $\mathcal{W}(I)$ plays the role of the energy.

Under our assumptions, the dual of $\ell\{M_B\}$ is isometrically isomorphic to $\ell\{M_B^*\}$ and the duality is expressed by the pairing

$$\langle I, J \rangle = \frac{1}{2} \sum_{B \in Y} i(B) j(B) \quad \text{where } I \in \ell\{M_B\} \text{ and } J \in \ell\{M_B^*\}.$$

Proposition (A.4). *Let I belong to $\ell\{M_B\}$. Then, the 1–chain H such that $h_B = p_B(i(B))$ belongs to $\ell\{M_B^*\}$.*

Taking Proposition (A.4) into account we may define the resistance operator from $\ell\{M_B\}$ to $\ell\{M_B^*\}$.

DEFINITION. The resistance operator $\mathcal{R}$ is defined by the following equation

$$\mathcal{R}(I) = \sum_{B \in X} p_B(i(B))\chi_B , \qquad I \in \ell\{M_B\}.$$

Note that in the linear case $\mathcal{R}$ coincide with the operator defined in Chapter I.

Let $\mathbf{Z}$ denote the linear subspace of *all* cycles in $\ell\{M_B\}$.

DEFINITION. We say that 1–chain $I \in \ell\{M_B\}$ satisfies Kirchhoff's equations if

(A.5) $\partial I + \imath \equiv 0$ (node law)

(A.6) $\langle Z, \mathcal{R}(I) \rangle = 0$ for all $Z \in \mathbf{Z}$ (loop law).

In particular, a 1–chain in $\ell\{M_B\}$ satisfying (A.5) is called a flow. A flow satisfying also (A.6) is called a current generated by $\imath$.

REMARK. In the linear case the current just defined coincides with the minimal current (compare with Theorem (3.21) and the Remark after the theorem).

The following result is the analogue of Theorem (3.25).

Proposition (A.7). *Let $(\Gamma, \mathcal{P})$ be a nonlinear network. Suppose that $\imath = \partial L$ for some $L \in \ell\{M_B\}$. Then, there exists a unique current $I \in \ell\{M_B\}$ generated by $\imath$, and I is the unique minimum point of $\mathcal{W}$ over all flows in $\ell\{M_B\}$.*

If I is the current dictated by Proposition (A.7), then by (A.6) there exists a unique (up to additive constants) function u on V such that

(A.8) $u(x) - u(y) = p_{[x,y]}(i(x,y))$ for all $x \sim y$.

Such a u is called the potential of I. It is easily seen that, if the functions q_B satisfy condition (A.1) as well, then the 1–chain $\{u(x) - u(y)\}_{[x,y] \in Y}$ given by (A.8) is in $\ell\{M_B^*\}$ if and only if $I \in \ell\{M_B\}$. In other words u satisfies

$$D(u) = \sum_{x \sim y} M_{[x,y]}^*(u(x) - u(y)) < \infty$$

if and only if $\mathcal{W}(I)$ is finite.

Arguing as in the linear case we can see that solving Kirchhoff's equations amounts to solving the nonlinear Poisson's equation

$$\Delta u + \imath = 0,$$

where Δ is the nonlinear Laplacian

(A.9) $$\Delta u = \sum_{y \sim x} q_{[x,y]}(u(x) - u(y)).$$

In the nonlinear case we clearly lose the the connection with Markov chain theory. However it is possible to define parabolic and hyperbolic networks and to mimic the proofs of some results of chapter III. This is still clearer in the case of p-networks.

2. p-networks

These networks are nonlinear networks such that

(A.10) $$p_B(i) = r(B)\text{sign}(i)|i|^{p-1}, \quad (1 < p < \infty),$$

The constants $r(B)$ are assigned positive "weights", which, in the linear case $p = 2$, can be interpreted as resistances of a linear network.

Yamasaki and his collaborators studied p-networks in great detail: we refer to the Bibliography for the papers containing the results described in this section.

The modular sequence space $\ell\{M_B\}$ in this case is nothing else but the weighted ℓ^p space (with weights $r(B)/p$). Hence we have the following definition.

DEFINITION. Let $1 < p < \infty$. We say that I has finite p-energy if

$$W_p(I) = \frac{1}{2} \sum_{B \in Y} r(B)|i(B)|^p < \infty.$$

The space of all 1-chains with finite p-energy is denoted by $\mathbf{H}_1^p$. The norm in $\mathbf{H}_1^p$ is obviously $W_p(I)^{1/p}$.

DEFINITION. Let u be a real valued function on V. We say that u has finite q-Dirichlet sum ($1 < q < \infty$) if, setting $c(x,y) = 1/r(x,y)$,

$$D_q(u) = \frac{1}{2} \sum_{y \sim x} c(x,y)^{q-1} |u(x) - u(y)|^q < \infty.$$

The space of all such functions with norm

$$||u|| = (|u(o)|^q + D_q)^{1/q},$$

where q is a reference vertex, is denoted by $\mathbf{D}_q$. As in the linear case we define the space $\mathbf{D}_{q,0}$ as the closure of ℓ_0 in $\mathbf{D}_q$.

Since the dual space of ℓ^p is ℓ^q (with p and q conjugate exponents), if u and I are related as in (A.8), then $I \in \mathbf{H}_1^p$ if and only if $u \in \mathbf{D}_q$.

The Laplacian (A.9) of the network takes the form

$$\Delta_q u(x) = \frac{1}{2} \sum_{y \sim x} c(x,y)^{q-1} \text{sign}(u(x) - u(y))|u(x) - u(y)|^{q-1}.$$

The notion of q–harmonic (q–superharmonic, q–subharmonic) function is defined by the relation $\Delta_q u = 0$ (≥ 0, ≤ 0). All these definitions become the usual ones for $p = q = 2$.

It is also possible to prove the analogue of Royden's decomposition theorem for the space $\mathbf{D}_q$.

As noted above, transience and recurrence do not make sense in this context, but we may define p–parabolic and p–hyperbolic networks.

DEFINITION. We say that a network is q–parabolic if

$$\inf\{D_q(u) : u \in \ell_0, u(a) = 1\} = 0 \quad \text{for some node } a.$$

Otherwise we say that the network is hyperbolic. By Theorem (3.41) a network is 2–parabolic if and only if it is recurrent.

Assume that we are given a graph Γ and weights $r(B)$ on the edges of Γ. For every fixed p ($1 < p < \infty$) we may define on Γ a structure of p–network (in particular for $p = 2$). Therefore the pair $(\Gamma, \{r(B)\})$ may be considered a p–network for all p.

In general, $(\Gamma, \{r(B)\})$ will be q–hyperbolic for certain values of p, and q–parabolic for other values of p (with p and q conjugate exponents). It is also clear that if $q_1 < q_2$, then q_1–parabolicity implies q_2–parabolicity. Thus we have the following definition.

DEFINITION. The parabolic index of a network (with respect to fixed weights) is the infimum of all q such that the network is q–parabolic. If the network is q–hyperbolic for all q, then we say that the parabolic index is ∞.

For instance Maeda (1977) proved that the parabolic index of $\mathbf{Z}^n$ is exactly n (see also Soardi and Yamasaki (1993)(b)). In the case of a homogeneous tree of degree greater than 2 it is not difficult to see that the parabolic index is ∞.

Many result of chapter III go through for p–networks. For instance, a necessary and sufficient condition for a network to be q–hyperbolic is that there exists a 1–chain $I \in \mathbf{H}_1^p$ such that $\partial I + \delta_a = 0$ (compare with Theorem (3.33); see also Kaimanovic (1992)). A version of Rayleigh's and Dirichlet's principles can also be proved for p–networks (and more generally for the nonlinear networks of §1; see Soardi (1993)(b)).

Sections 7–9 of chapter III hold almost entirely for nonlinear p–networks. Uniqueness is discussed in Soardi and Yamasaki (1993), along the lines of Theorem (4.8) in chapter IV. The definition and the first steps in the study of p–Royden's boundary are due to Yamasaki (1987) and Kayano and Yamasaki (1988). Soardi and Yamasaki (1993)(b) prove that the parabolic index is invariant by rough isometries.

REMARK. As in the linear case, the theory of p–network is the discrete counterpart of a continuous theory; see for example Lindqvist (1986), Sakai (1994) and the bibliography therein contained.

A nonlinear potential theory connected with the quasilinear elliptic equations on Riemannian manifolds

$$-\mathrm{div}\mathcal{A}_x(\nabla u) = 0$$

has been developped in the recent years especially by the finnish school (we refer to Holopainen (1990) and the bibliography therein contained). It would be interesting to know whether the results outlined in §1 of this Appendix may be considered the discrete analogues of such a theory.

BIBLIOGRAPHY

L. Affer (1992), *Reti elettriche infinite non localmente finite*, Thesis, Università di Milano.

A. Ancona (1988), *Positive harmonic functions and hyperbolicity*, Potential Theory, Surveys and Problems. Springer Lecture Notes in Mathemathics n. 1344, eds. J.Král et al., pp. 1–23.

A. Ancona (1990), *Théorie du potential sur les graphes et les variétés*, École de probabilité de St. Flour Springer Lecture Notes in Mathemathics n. 1427, pp. 1–311.

A.F. Beardon (1983), *The Geometry of Discrete Groups*, Springer Verlag, New York, Berlin, Heidelberg.

M.E.B. Bekka and A. Valette (1994), *Group cohomology, harmonic functions and the first L^2 Betti number*, preprint.

I. Benjamini and Y. Peres (1992), *Random walks on a tree and capacity in the interval*, Ann. Inst. Henri Poincaré **28**, 557-592.

A. Beurling and J. Deny (1959), *Dirichlet spaces*, Proc. Nat. Acad. Sci. U.S.A. **45**, 208-215.

B. Bollobás (1979), *Graph theory: an introductory course*, Springer Verlag, Berlin Heidelberg New York.

M. Breen (1985), *Tilings whose members have finitely many neighbors*, Israel J. Math. **52**, 140-146.

R. Brooks (1985), *The spectral geometry of the Apollonian packing*, Comm. Pure Appl. Math. **38**, 357-366.

L. Carleson (1967), *Selected problems on exceptional sets*, Van Nostrand, Princeton, New Jersey.

P. Cartier (1972), *Fonctions harmoniques sur un arbre*, Symp. Math. **9**, 203-270.

D.I. Cartwright and W.Woess (1992), *Infinite graphs with nonconstant Dirichlet finite harmonic functions*, SIAM J. Discrete Math. **3**, 380-385.

D.I. Cartwright, P.M. Soardi and W.Woess (1993), *Martin and end compactifications for non locally finite graphs*, Trans. Amer. Math. Soc. **338**, 679-693.

C. Cattaneo (1994), *Trees which satisfy the strong isoperimetric inequality with weights*, Boll. Un. Mat. Ital. **8-A**, 75-82.

C. Constantinescu (1962), *Dirichletsche Abbildungen*, Nagoya Math. J. **20**, 75-89.

C. Costantinescu and A. Cornea (1963), *Ideale Ränder Riemannschen Flächen*, Springer Verlag, Berlin, Heidelberg, New York.

Th. Coulhon and L. Saloff-Coste (1994), *Variété riemanniennes isométriques à l'infinie* (to appear).

R. Courant, K. Friedrichs and H. Lewy (1928), *Über die partiellen Differerenzengleichungen der mathematischen Physik*, Math. Ann. **100**, 32-74.

L. De Michele and P.M. Soardi (1990), *A Thomson's principle for infinite, nonlinear, resistive networks*, Proc. Amer. Math. Soc. **109**, 461-468.

J. Dodziuk (1984), *Isoperimetric inequality and transience of certain random walks*, Trans. Amer. Math. Soc. **284**, 787-794.

J.L. Doob(1984), *Potential theory and its probabilistic counterpart*, Springer Verlag, Berlin Heidelberg New York.

P.G. Doyle (1988), *Electric currents in infinite networks*, unpublished manuscript.

P.G. Doyle and J.L. Snell (1984), *Random walks and electric networks*, Mathematical Association of America, Washington D.C., 1984.

R.J. Duffin (1947), *Nonlinear networks*, Bull. Amer. Math. Soc. **53**, 963–971.

R.J. Duffin (1953), *Discrete potential theory*, Duke Math. J. **20**, 233–251.

R.J. Duffin (1962), *The extremal length of a network*, J. Math. Anal. Appl. **5**, 200–215.

R.J. Duffin and D.H. Shaffer (1960), *Asymptotic expansion of double Fourier transforms*, Duke Math. J. **27**, 581–596.

J. Fabrykowski and N. Gupta (1985), *On groups with subexponential growth function*, J. Indian Math. Soc. **49**, 249–256.

A. Figà-Talamanca and M. Picardello(1983), *Harmonic analysis on free groups*, M. Dekker, New York.

H. Flanders (1971), *Infinite networks I - Resistive networks*, IEEE Trans. Circuit Theory **CT-18**, 326–331.

H. Flanders (1972), *Infinite networks II - Resistance in an infinite grid*, J. Math. Anal. Appl. **40**, 30–35.

H. Flanders (1974), *A new proof of R. Foster's averaging formula in networks*, Linear Algebra and its Appl. **8**, 35–37.

R.M. Foster (1949), *The average impedance of an electrical network*, Contributions to applied Mechanics, Reissner Anniv. Vol., Edwards Bros., Ann Arbor, pp. 333–340.

H. Freudenthal (1944), *Über die Enden diskreter Räume und Gruppen*, Comm. Math. Helv. **17**, 1–38.

M. Fukushima (1980), *Dirichlet forms and Markov processes*, North Holland, Amsterdam, Oxford, New York.

P. Gerl (1986), *Eine isoperimetrische Eigenschaft von Bäumen*, Sitzungsber. Öst. Akad. Wiss. Math. Naturw. Kl. **195**, 49–52.

P. Gerl (1988), *Random walks on a graph with a strong isoperimetric property*, J. Theoretical Prob. **1**, 171–188.

M. Glasner and R. Katz (1970), *The Royden boundary of a Riemannian manifold*, Illinois J. Math. **14**, 488–495.

M. Glasner and R. Katz (1982), *Limits of Dirichlet finite functions along curves*, Rocky Mountain J. Math. **12**, 429–435.

R.I. Grigorchuk (1983), *On Milnor's problem of group growth*, Soviet Math. Dokl. **28**, 23–26.

R.I. Grigorchuk (1985), *Degrees of growth of finitely generated groups and the theory of invariant means*, Math. USSR Izvestiya **25**, 259–300.

A.A. Grigor'yan (1988), *On Liouville theorems for harmonic functions with finite Dirichlet integral*, Math. USSR Sbornik **60**, 485–504.

M. Gromov (1981), *Hyperbolic manifolds, groups and actions*, Riemann surfaces and related topics: Proceedings of the 1978 Stony Brook conference, ed. I. Kra and B. Maskit, Princeton Univ. Press, Princeton, New Jersey, pp. 183–213.

M. Gromov (1987), *Hyperbolic groups*, Essays in group theory, ed. S.M. Gersten, Springer Verlag, New York, pp. 75–263.

B. Grünbaum and G.C. Shephard (1987), *Tilings and patterns*, Freeman, New York.

R. Halin (1964), *Über unendliche Wege in Graphen*, Math. Ann. **157**, 125–137.

R. Halin (1973), *Endomorphisms and automorphisms of infinite locally finite graphs*, Abh. Math. Sem. Univ. Hamburg **39**, 251–283.

I. Holopainen (1990), *Nonlinear potential theory and quasiregular mappings on Riemannian manifolds*, Ann. Acad. Sci. Fenn., Ser A I Math. Dissertationes **74**, 1–45.

I. Holopainen (1994), *Rough isometries and p-harmonic functions with finite Dirichlet integral*, Revista Matemática Iberoamericana **10**, 143–176.

W. Imrich and N. Seifter (1988/89), *A note on the growth of transitive graphs*, Discrete Math. **73**, 111–117.

W. Imrich and N. Seifter (1991), *A survey on graphs with polynomial growth*, Discrete Math., 101–117.

V.A. Kaimanovic (1992), *Dirichlet norms, capacities and generalized isoperimetric inequalities for Markov operators*, Potential Anal. **1**, 61–82.

S.L. Kalpazidou (1991), *On Beurling inequality in terms of thermal power*, J. Appl. Prob. **28**, 104–115.

S.L. Kalpazidou (1995), *Cycle representations of Markov processes*, Springer Verlag, Berlin, New York, Heidelberg.

M. Kanai (1985), *Rough isometries and combinatorial approximation of geometries of non-compact Riemannian manifolds*, J. Math. Soc. Japan **37**, 391–413.

M. Kanai (1986)(a), *Rough isometries and the parabolicity of Riemannian manifolds*, J. Math. Soc. Japan **38**, 227–238.

M. Kanai (1986)(b), *Analytic inequalities and rough isometries between noncompact Riemann manifolds*, Curvature and topology of Riemannian manifolds. Springer Lecture Notes in Mathematics 1201, ed K. Shiohama, T. Sakai and T. Sunada, pp. 122–137.

M.I. Kargapolov and Ju.I. Merzljakov (1979), *Fundamentals of the theory of groups*, Springer Verlag, New York, Berlin, Heidelberg.

T. Kayano and M. Yamasaki (1984), *Boundary limits of discrete Dirichlet potentials*, Hiroshima Math. J. **14**, 401–406.

T. Kayano and M. Yamasaki (1988), *Some properties of Royden boundary of an infinite network*, Mem. Fac. Sci. Shimane Univ. **22**, 11–19.

J.G. Kemeny, J.L. Snell, A.W. Knapp (1966), *Denumerable Markov chains*, Van Nostrand, Princeton, New Jersey.

M. Konsowa (1992), *Effective resistance and random walks on finite graphs*, Arabian J. Sci. Engrg. **17**, 181–184.

G. Kuhn (1991), *Finitely additive random walks on infinitely generated free groups*, J. Theoret. Prob. **4**, 311–320.

Y. Kusunoki and S. Mori (1959), *On the harmonic boundary of an open Riemann surface, I*, Japan. J. Math. **29**, 52–56.

Y. Kusunoki and S. Mori (1959), *On the harmonic boundary of an open Riemann surface, II*, Mem. Coll. Sci. Kyoto Ser. A Math. **33**, 209–233.

R. Larsen (1973), *Banach Algebras*, M. Dekker, New York.

P. Lindqvist, *On the definition and properties of p–superharmonic functions*, J. Reine Angew. Math. **365**, 67–79.

R. Lyons (1993), *Random walks and percolation on trees*, Ann. of Probability (to appear).

T. Lyons (1983), *A simple criterion for transience of a reversible Markov chain*, Ann. of Probability **11**, 393–402.

F-Y. Maeda (1977) A remark on the parabolic index of infinite networks, Hiroshima J. Math. **7**, 147–152.

W. Magnus (1974), *Noneuclidean tesselations and their groups*, Academic Press, New York and London.

S. Markvorsen, S. McGuinness and C. Thomassen (1992), *Transient random walks on graphs and metric spaces with applications to hyperbolic surfaces*, Proc. London Math. Soc. **64**, 1–20.

W.H. Mc Crea and F.J.W. Whipple (1940), *Random paths in two and three dimensions*, Proc. Roy. Soc. Edinburgh **60**, 281–298.

S. McGuinness (1991), *Recurrent networks and a theorem of Nash–Williams*, J. Theoretical Prob. **4**, 87–100.

G. Medolla and P.M. Soardi (1993), *Extension of Foster's averaging formula to infinite networks with moderate growth*, to appear in Math. Z..

G. Medolla (1994), *Rough isometries between networks* (to appear).

B. Mohar (1988), *Isoperimetric inequalities, growth and the spectrum of graphs*, Linear Alg. and its Appl. **103**, 119–131.

M. Murakami and M. Yamasaki (1992), *Nonlinear potentials on an infinite network*, Mem. Fac. Sci. Shimane Univ. **26**, 15–28.

M. Nakai (1994), *Harmonic measure on euclidean balls*, Nagoya Math. J. **133**, 85–126.

T. Nakamura and M. Yamasaki (1977), *Extremal length of an infinite network which is not necessarily locally finite*, Hiroshima Math. J. **7**, 813–826.

C.St.J.A. Nash-Williams (1959), *Random walks and electric currents in networks*, Proc. Cambridge Phil. Soc. **55**, 181–194.

A. Nevo (1991), *A structure theorem for boundary transitive graphs*, Israel J. Math. **75**, 1–20.

P. Pansu (1989), *Cohomologie L^p des variétés à courbure négative, cas du degré 1*, Fascicolo Speciale, Rend. Sem. Mat. Univ. Pol. Torino, 95–120.

W. Paschke (1993)(a), *The flows space of a directed graph*, Pacific J. Math. **159**, 127–138.

W. Paschke (1993)(b), *A numerical invariant for finitely generated groups via action on graphs*, Math. Scand. **72**, 148–166.

G. Polya (1921), *Über eine Aufgabe der Wahrscheinlichkeitrechnung betreffend die Irrfahrt im Strassennetz*, Math. Ann. **84**, 149–160.

H.L. Royden (1952), *Harmonic functions on open Riemann surfaces*, Trans. Amer. Math. Soc. **73**, 40–94.

H.L. Royden (1953), *On the ideal boundary of a Riemann surface*, Annals of Mathematics studies, Princeton **30**, 107–109.

G. Sabidussi (1964), *Vertex transitive graphs*, Monatsh. Math. **68**, 427– 438.

M. Salvatori (1992), *On the norms of group–invariant transition operators on graphs*, J. Theoretical Prob. **5**, 563–576.

L. Sario and M. Nakai (1970), *Classification theory of Riemann surfaces*, Springer Verlag, Berlin, Heidelberg, New York.

E. Schlesinger (1992), *Infinite networks and Markov chains*, Boll. Un. Mat. Ital. **6–B**, 23–37.

N. Seifter (1991), *Properties of graphs with polynomial growth*, J. Combinatorial Theory Series B **52**, 222-235.

H. Shapiro (1987), *An electric lemma*, Math. Magazine **60**, 36-39.

C.L. Siegel (1971), *Topics in complex function theory, vol. II*, Wiley-Interscience, New York, London, Sydney.

V.Ya. Skorobogat'ko (1972), *A test for the convergence of a branching continued fraction*, Dopovidi Akad. Nauk Ukrain.-RSR Ser. A **93**, 27–29.

P.M. Soardi (1990), *Recurrence and transience of the edge graph of a tiling of the euclidean plane*, Math. Ann. **287**, 613–626.

P.M. Soardi (1993)(a), *Rough isometries and Dirichlet finite harmonic functions on graphs*, Proc. Amer. Math. Soc. **119**, 1239–1248.

P.M. Soardi (1993)(b), *Morphisms and curents in infinite nonlinear resistive networks*, Potential Anal. **2**, 315–347.

P.M. Soardi and W. Woess (1990), *Amenability, unimodularity, and the spectral radius of infinite graphs*, Math. Z. **205**, 471–486.

P.M. Soardi and W. Woess (1993), *Uniqueness of currents in infinite resistive networks*, Discrete Appl. Math. **31**, 37–49.

P.M. Soardi and M. Yamasaki (1993), *Classification of infinite networks and its application*, Circuits Systems Signal Process. **12**, 133–149.

P.M. Soardi and M. Yamasaki (1993)(b), *Parabolic index and rough isometries*, Hiroshima J. Math **23**, 333–342.

F. Spitzer (1964), *Principles of random walks*, Van Nostrand, Princeton.

A. Stöhr (1950), *Über enige lineare partielle Differenzengleichungen mit konstanten koeffizienten I, III*, Math. Nach. **3**, 202–242 and 330–357.

P. Tetali (1991), *Random walks and the effective resistance of networks*, J. Theoretical Prob. **4**, 101–109.

P. Tetali (1994), *An extension of Foster's network theorem* Combinatorics, Probability and Computing, (to appear).

C. Thomassen (1989), *Transient random walks, harmonic functions, and electrical currents in infinite electrical networks*, Mat-Report n. 1989-07, The Technical Univ. of Denmark.

C. Thomassen (1990), *Resistances and currents in infinite electrical networks*, J. Combinatorial Theory Ser. B **49**, 87–102.

V.I. Trofimov (1984), *Automorphisms of graphs and a characterization of lattices*, Math. USSR Sbornik **22**, 379–391.

V.I. Trofimov (1985)(a), *Graphs with polynomial growth*, Math. USSR Sbornik 51, 405–417.

V.I. Trofimov (1985)(b), *Automorphisms group of graphs as topological group*, Math. Notes **38**, 717–720.

M. Tsuji (1975), *Potential theory in modern function theory*, Chelsa, New York.

B. Van der Pol (1959), *The finite difference analogue of the periodic wave equation and the potential equation*, Appendix IV in M. Kac "Probability and related topics in physical sciences" Interscience, London.

N.Th. Varopoulos (1985), *Isoperimetric inequalities and Markov chains*, J. Funct. Anal. **63**, 215–239.

W. Woess (1994), *Random walks on infinite graphs and groups – a survey on selected topics*, Bull. London Math. Soc. **26**, 1–60.

J. Y. T. Woo (1973), *On modular sequence spaces*, Studia Math. **48**, 271–289.

J. Wysoczański (1994), *Royden compactification of integers*, preprint.

M. Yamasaki (1975), *Extremum problem on an infinite network*, Hiroshima Math. J. **5**, 223–250.

M. Yamasaki (1976), *Generalized extremal length of an infinite network*, Hiroshima J. Math. **6**, 95–111.

M. Yamasaki (1977), *Parabolic and hyperbolic infinite networks*, Hiroshima Math. J. **7**, 135–146.

M. Yamasaki (1979), *Discrete potential on an infinite network*, Mem. Fac. Sci. Shimane Univ. **13**, 31–44.

M. Yamasaki (1982), *Dirichlet finite solutions of Poisson equation on an infinite network*, Hiroshima Math. J. **12**, 569–579.

M. Yamasaki (1986), *Ideal limit of discrete Dirichlet functions*, Hiroshima Math. J. **16**, 353–360.

M. Yamasaki (1987), *Discrete Dirichlet potentials on an infinite network*, R.I.M.S Kokyuroku **610**, 51–66.

M. Yamasaki (1991), *Discrete initial value problem and parabolic potemtial theory*, Hiroshima Math. J. **21**, 285–299.

M. Yamasaki (1992), *Nonlinear Poisson equations on an infinite network*, Mem. Fac. Sci. Shimane Univ. **23**, 1–9.

K. Yosida (1970), *Functional analysis*, Springer Verlag, Berlin Heidelberg New York.

A.H. Zemanian (1976)(a), *Infinite electrical networks*, Proc. IEEE **64**, 6–17.

A.H. Zemanian (1976)(b), *The complete behavior of certain infinite networks under Kirchhoff's node and loop laws*, SIAM J. Appl. Math. **30**, 278–296.

A.H. Zemanian (1978), *The limb analysis of of countably infinite electrical networks*, J. Combinatorial Theory, Series B **24**, 76–93.

A.H. Zemanian (1982), *Nonuniform semi-infinite grounded grids*, SIAM J. Math. Anal. **13**, 770–788.

A.H. Zemanian (1991)(a), *Infinite electrical networks*, Cambridge University Press, Cambridge.

A.H. Zemanian (1991)(b), *An electrical gridlike structure excited at infinity*, Math. Control Signal Systems 4, 217–231.

A.H. Zemanian and P. Subramanian (1983), *A theory of ungrounded grids and its applications to the geophysical exploration of layered strata*, Studia Math. **77**, 162–181.

INDEX

Printing: Weihert-Druck GmbH, Darmstadt
Binding: Theo Gansert Buchbinderei GmbH, Weinheim

Lecture Notes in Mathematics

For information about Vols. 1–1411
please contact your bookseller or Springer-Verlag